Household Waste Management in Europe

Household Waste Management in Europe

economics and techniques

Edited by

A.V. Bridgwater
Department of Chemical Engineering
University of Aston in Birmingham, UK

K. Lidgren
The TEM Group
University of Lund, Sweden

VAN NOSTRAND REINHOLD COMPANY

New York — Cincinnati — Toronto — London — Melbourne

**Published by Van Nostrand Reinhold Company Ltd.,
Molly Millars Lane, Wokingham, Berkshire, England.**

*Published in 1981 by Van Nostrand Reinhold Company,
135 West 50th Street, New York, NY 10020, USA*

*Van Nostrand Reinhold Limited,
1410 Birchmount Road, Scarborough,
Ontario, M1P 2E7, Canada*

*Van Nostrand Reinhold Australia Pty. Limited,
17 Queen Street, Mitcham, Victoria 3132, Australia.*

Library of Congress Cataloging in Publication Data

Main entry under title:

Household waste management in Europe.

Includes bibliographies and index.
1. Refuse and refuse disposal—Europe.
I. Bridgwater, A. V. II. Lidgren, Karl, 1947-
TD789.E9H68 363,7'28 81-10439
ISBN 0-442-30464-1 AACR2

Printed and bound in Great Britain
at The Pitman Press, Bath

Foreword

Waste management plays an increasingly important role in the overall debate on the protection of our environment and the rational use of raw materials. This has become quite clear during the last ten years, when the waste problem has received public attention at all levels. Politicians, the press and the public in general recognize the need for proper waste management, as do experts from industry and from the scientific sector.

In contrast to the general consensus that something should be done, there are often fundamentally different views on what could and should be done to solve the waste problem in both technical and economic terms. Many discussions in the past varied between too optimistic expectations on the one side and helplessness on the other, when being asked how to introduce a new waste management concept.

This picture is now changing. Fortunately vague generalizations on waste disposal and waste recycling, which often characterized the discussion in the past, are being replaced by a more channeled and realistic evaluation. Today, we are aware of the complexity of the problem and the impossibility of offering general solutions to it. There is no common recipe which fits each individual waste situation or commodity. What we need are detailed studies and evaluations of special technical attributes and of the particular requirements involved in collection and treatment of individual waste types and, last but not least, a good understanding of the secondary materials markets.

In this respect, this book on household waste management is an excellent contribution to improve our knowledge and to make us familiar with the international state of the art in the field. Written by outstanding authors from academia and waste management practice, the book makes clear what the actual problems are; where the potential for present and future improvements and measures is; and which technical and economic limitations exist for all our attempts to get the household waste problem under control. The individual chapters confirm that household waste disposal and recycling are not isolated but integrated tasks — a message that cannot be emphasized strongly enough for all who take over responsibility in the field.

Household waste management has made considerable progress in the last decade. Nevertheless, many activities, in particular in the area of resource recovery from household wastes, are only at an early stage of development. The lack of practical experience and comprehensive information is obvious. This book helps essentially to full this gap and to put the problem in the right perspective. This is what we need to make further progress.

Dr. Bernd Wolbeck

Contents

1 The Waste Problem

Karl Lidgren

INTRODUCTION

Time and time again, history has proved how new problems that have arisen have initially been treated with ignorance. However, as knowledge and experience have increased and been coupled with new ideas and methods, the problems have often been overcome and even eliminated. We believe that this simple observation is worth developing in relation to one of our current greatest and most debated problems: the risk of over-utilization of our environment. (In this case 'environment' has considerable scope, encompassing both the material and the energy consumption that affect the external environment in the form of water, air and ground pollution.)

No one today can definitely say how great — or how small — the risk really is that man's way of living will seriously damage nature, and thereby his own opportunities in life. However the one thing that is certain for example is the concern regarding ecological risks which has reached a wide public, for example in the form of demonstrations against nuclear power and the formation of ecological organizations.

The continuous stream of alarming threats to the environment made by the media has raised the temperature of debate and provoked political intervention behind which one can glimpse both political tactics and positive arguments. However, although the media are indisputably useful — superficially yet effectively — in taking up the problems and querying actions, the weight of the media puts pressure on the political system to act in a way that does not always produce the answer to the real needs or the best long-term solution. Accumulated knowledge and cautious problem evaluation easily become secondary questions in this escalating war of influence and public opinion.

Questions that directly affect everyday life, such as disposable containers, receive interest out of all proportion to their importance, from both the media and the authorities. Unfortunately, some politicians knowingly exploit this distortion towards less dangerous products. It is said that it is educationally and psychologically important to impress upon the citizen; in this way, the government's aims regarding the environment, even if these measures have very little effect (which, is shown in an example in this chapter) can even have an overall negative effect.

With these phenomena in society's debates and goverments there is a definite risk that interest, investigations, opinion forming and political governing power will be aimed consistently at the most dangerous and most important questions but will lead to a number of detailed regulations to emphasize 'the government's environmental political awareness'. To counterbalance the part played by the tactical dispositions of the public, there are deeper studies and investigations carried out by government offices and parliamentary commissions. Here scientific systematized knowledge definitely has a better chance of succeeding. So far it has not been very easy to bring about a fruitful meeting between this scientific knowledge, the political conditions and the real problems. This may be because the scientific studies used are to a greater or lesser extent, very limited in scope and depend among other things on the research scientist's specialization and aim and the grants at his disposal. As a rule, the political directives for such investigation do not deal with a complete perspective but introduce further limitations, often based on an acute and urgent political problem. It is therefore not surprising that when the resultant scientific documents meet restricted political directives it is difficult to arrive at the correct conclusions to the question concerned. It also means that next time society will give even less priority to the matter that should be the most important in ecological politics.

During the first intensive decades of environmental research, many of the studies typically concerned a definite and limited problem, in which it is scientifically fundamental to collect all the relevant facts. However, the work has now reached the point when there is good reason, from the research scientist's point of view, to pay more attention to synthesis and total perspective, to make it easier for politicians to carry out their evaluations in a correspondingly wider perspective of priorities.

It is in this spirit that this book about disposal problems presents the results from a week-long seminar for ecologists from all over Europe. We hope and believe that our report will answer certain questions that are of particular interest to various specialists. But we also hope that our attempt to present the subject of household waste as part of a much greater politics, economic efficiency and inefficiency, and positive environmental impact, including any negative side effects, for society in general.

During the last decade the industrialized countries have been subjected to considerable price rises for raw materials and energy. The ever-increasing 'waste mountain' has, by many, been taken as a gross indication of the misuse of these valuable raw materials. Criticism has in particular been directed against the disposal method of landfilling. The arguments here are that this disposal method hardly constitutes a meaningful final stage of a process in which unprocessed raw materials are dug up, refined, used in the form of products, and finally replaced in the earth in the form of refined raw materials.

Growing material living standards have also led to an increased demand for an improved external environment. Many waste treatment processes have a negative effect on the environment in the form of water and air pollution and poor aesthetic effects. In many cases there is therefore a conflict between the demand for an acceptable environment and the demand that the waste be taken care of in some way.

To begin with, one way of approaching the 'waste problem' could be to try to answer two questions:

(i) How important is household waste as an environmental risk?
(ii) How important is household waste where the possibilities of reducing the consumption of raw materials that are in short supply are concerned?

HOUSEHOLD WASTE AS AN ENVIRONMENTAL RISK

At present it is not possible to draw up a list ranking all the risk factors due to lack of knowledge, such as unknown side effects. Nonetheless, it is valuable roughly to classify grades of risk, based on presently available knowledge to stimulate debate.

One way of classifying these environmental risks is as follows:

Class 1: obviously poisonous or otherwise dangerous products which, when spread in an uncontrolled manner, can result in immediate injuries to human beings and the environment, e.g.

Nuclear fuel processes and their waste products, chemical processes, products and transport, and chemical waste products.

The quantity of 'poisons and hazardous waste' now dealt with in the Common Market is estimated at more than 20 million tonnes, compared with 85-90 million tonnes of household waste arising in the same area.

Class 2: gradual environmental risks that slowly and visibly threaten humanity's and the environment's wellbeing and their capacity to restore good health. These include known risk factors whose real and potential destructive capacities are not known exactly and which perhaps cannot be evaluated in relation to one another. However, it should be possible to differentiate between the more serious and the less serious. Examples from this class include exhaust from motor vehicles, fumes from combustion processes, over-exploitation and pollution of ground water, acidification of lakes and forests, and use of intractable poisons in agriculture, forestry and the food industry.

Class 3: negative aesthetic environmental effects (such as changes that do not threaten the biological environment's ecology but can be said to deteriorate natural environmental assets), e.g. building of recreation grounds, badly sited factories chimney stacks, motorways and refuse dumps, litter in the country side.

If we are going to assess household waste from the viewpoint of class of risk, we should primarily include it in the risk-free aesthetic Class 3, or state that it does not generally involve any of the more serious risks to the environment of Classes 1 and 2. However, household waste includes many components and a few of these can be placed in Class 1 and should therefore be specially dealt with. These include chemical residues such as insecticides, remains of paint and solvents, medicines and cadmium batteries. It should be mentioned that older badly organized or managed refuse dumps are also an ecological hazard according to Class 2 in the form of leachate which can pollute ground water. Such problems are usually carefully controlled in modern disposal methods with known techniques.

The greatest disadvantage of household waste is that disposal requires space, and locally may contribute to the aesthetically negative elements in the environment, although normally waste disposal can be arranged in suitable places. There is no real lack of space for waste disposal, but it may (for aesthetic environmental reasons) be desirable to place the disposal sites at a greater, and consequently more expensive, distance from the areas where the waste is produced.

During recent years, different techniques have been tested to utilize waste as raw materials. However, some of these techniques can give rise to environmental risks according to Class 2, for example in the form of air pollution from incineration.

In summary, the following conclusion may be drawn concerning household waste as a risk to the environment. The ecological hazards according to Classes 1 and 2 must generally be considered as the risks to which society should give priority. Compared with the wide range of risk factors that can be encountered, household waste only contains marginal amounts of risk products that belong to these classes. Household waste may therefore be primarily regarded as a volumetric problem requiring attention, rather than an enviromental risk problem.

HOUSEHOLD WASTE AND REDUCTION OF CONSUMPTION OF RAW MATERIALS

The concept that, as far as it is theoretically possible, one should try to utilize all raw materials from household waste has become quite common. Sometimes this attitude is based on the understanding that by means of such utilization it might be about to contribute to the world's decreasing access to raw materials.

The conformity is not good when comparing a list of the raw materials considered to be the most threatened with those that can be recovered from household waste. The most significant bulk problem in household waste is paper, and paper comes from a renewable source — the forest. Certainly the demand on this resource is great, but there are substantial possibilities for expanding this potential. Similarly recovery of glass, which has been demanded by authorities in many countries cannot be justified by a lack of raw materials either — there is an abundance of sand and soda.

Furthermore, there is good reason to check the term 'shortage' more closely and more critically. Shortage is most frequently a relative, not an absolute, term in the developed countries. In the last century it was said that the world stood on the brink of a catastrophe because the supply of coal was about to end. Reality proved very definitely that this prediction was wrong, instead coal is a present day and future resource for adapting to a new energy and raw materials future in the 21st century. When the shortage is a temporary difference between a great demand and a somewhat lower production, and where the possibilities of a transition to a substitute are good, the shortage is only a relative market term, which should not necessarily be taken as a justification for (as in the case of household waste) controls of very marginal importance.

Finally, it must be questioned whether the recovery of waste would reduce the pressure on the supply of raw materials: the great demand for raw materials does not depend on the fact that part of them become household waste, but on the fact that the economic welfare has so far been built on the expanding production and consumption of raw materials. Just as recovery can reduce the pressure on the supply of raw materials, it can also be used as a contribution towards an increasing economy. In the formation of public opinion the actual recovery can be seen as a justification of continued intensive exploitation of raw materials.

So far we have come to the conclusions that:

(i) in spite of the occurrence of certain products in household waste that are dangerous to the environment, this does not necessarily mean that it is a serious ecological hazard;

(ii) in spite of the fact that household waste should be treated as a resource, the recovery of this waste should not necessarily be accepted as a solution to the world's raw material problem.

Even if household waste as such is neither a great threat to humanity (as an ecological hazard) nor a great chance for humanity (as a new source of raw materials) the fact remains that waste is produced and creates a bulk problem. The question which is now natural to ask and attempt to answer is how will waste volumes develop in the future?

FUTURE DEVELOPMENT OF WASTE VOLUME

In a highly industrialized country where production as well as consumption is high, a large amount of waste is generated as a result of both activities. In consumer-originating waste, newspapers and packaging are examples of products of very short life, whereas irons, vacuum cleaners and refrigerators are goods of a considerably longer life expectancy. A feature common to them all however, is, that they will end up as waste sooner or later.

There are in fact two main sources of waste:

the production system,
the consumption system.

Most companies in all sectors of industry are striving for an increased degree of internal recovery, and attempts are continually being made to increase the degree of external recovery, i.e. recovery of what is normally referred to as waste.

A question that must be asked in this connection is, therefore, whether quantities of waste will decrease or increase in the future. An increase in the degree of internal recovery will, naturally, lead to a decrease, calculated as a percentage, in the waste products from the production system. But this need not lead to a decrease in the total amount of waste. As was mentioned earlier, waste is generated both from the production system and from the consumption system. Even if internal recovery becomes more efficient, the quantity of waste from the production system could increase all the same, as a result, for example, of an increase in production. The quantity of waste from the consumption system is not directly affected by a change in the internal degree of recovery.

Essential to the amount of waste generated is also the quantity of goods consumed. A smaller quantity of waste can be achieved either by reducing consumption or by using products of longer life. Recovery has often been considered the opposite of the principle buy-use-throw away, but the fact is that if there were a radical deviation from this principle, the lack of waste would reduce the justification for external recovery. A reasonable answer to the question raised above will therefore be that the quantities of waste are not likely to diminish in future.

The premises for this book can be summarized from the discussion above as follows:

(i) From a general point of view, household waste does not constitute an environmental hazard.
(ii) Waste is a resource, but as such should not be overvalued.
(iii) The quantities of solid waste are not likely to decrease in future and it will therefore continue to be important to look at the costs within the waste disposal sector, as within any other sector.
(iv) The real costs for raw materials and energy are not likely to decrease in future.

(v) There will be increased demands for a better environment.

The contributors to this book are researchers from different European countries. They have looked at the problems connected with waste disposal from different starting-points, which is quite natural given their different geographic domiciles and their different reseach interests — some of them being economists, other technologists. Common to all authors is, however, an awareness that the premises listed above are essential underlying factors for the so-called waste problem, and the findings of each contribution provide substantive support for these views. The justification for the work is, however, more usually derived from the need to solve a particular problem — and there are a large number of questions to be answered.

SOME COMMON QUESTIONS ENCOUNTERED CONCERNING THE WASTE PROBLEM

In public debate, the waste problem has been surrounded by a great number of specific questions. Some of the more common questions put below are answered, directly or indirectly, in the following chapters. The concluding chapter will revert more explicitly to the answers to these questions.

Question 1

In many countries arguments have been put forward, mainly from political quarters, for a variety of interventions, which would result in a finite reduction of the waste volume. This seems very attractive as waste management costs are increasing considerably in most countries. The natural question is: can costs be reduced by reducing the waste volume by, say, 5-10%?

Question 2

If the beverage container debate that has continued in many European countries, as well as in the USA, during the last decade is analysed the impression is obtained that if we would only change to refillable bottles for certain beverages the waste problem would be solved, and hence also a number of other problems. Considering the modest contribution of beverage containers to the total waste quantity, it should be asked if this is really true and in either case whether the debate on different packaging systems for beverages has been given disproportionate attention in mass media and in general debate.

Question 3

One way of making producers as well as consumers pay more attention to the generation of waste may be to let some kind of waste treatment charge be included in the price of the product. Were this possible, the signals of the price system would be passed both to producer and consumer. Is such a charge system realistic and practicable?

Question 4

A large number of proposals have been made recently as to how wastes can be taken care of in an ecologically and economically acceptable way, by processing them in large plants. A natural question is therefore: should our society not invest more resources in plants for central sorting of the wastes?

Question 5

In many ways, waste management and recovery can be said to be a question of cost. As the production of waste is generally considered to create problems, and as recovery is generally considered to be the solution to these problems, there is good reason to ask whether society should not invest money in encouraging those waste-management and recovery activities that are considered desirable.

Question 6

The market for recovered material is, to say the least, fairly uncertain. Prices fluctuate considerably and the recovered material often serves as a buffer which is used during a boom, but not during an economic recession when natural raw materials are used instead. If investing in a recovery scheme it is natural to expect, or at least prefer, stable conditions in much the same way as for any other industrial activity. What can be done to achieve market conditions that will lead to an acceptable degree of recovery?

Question 7

In the discussion of different recovery activities, energy balances are often presented to justify taking a certain action. Should decisions of this kind be made on the basis of these energy balances alone?

Question 8

There are those who claim that the waste problem will not really be solved until there is a change in attitudes of producers and consumers. Producers must be made to produce goods that generate less waste, and, in addition, the waste generated should be easy to recycle or dispose of. Consumers must become more observant of their own waste production and should be prepared to aid more effective waste management, for example by assisting in some sorting work concerning products like newspapers. Can this change in attitude be achieved in a simple way?

Question 9

For combustible material, such as paper, there is sometimes a 'recovery conflict' between on the one hand material recovery, and on the other hand energy recovery by combustion of the waste. The argument is sometimes postulated that paper can only be burned once, whereas it can be used over and over again through material recovery. This argument is sometimes used to rank material recovery above energy recovery as a recycling method. The question is: is material recovery always preferable to energy recovery?

Question 10

From different quarters, attention has been drawn to the suggestion that the main reason for the so-called waste problem is that it is nearly impossible to find new dumping sites. Is this really so, and to what consequences does this lead?

Some of these questions are dealt with in the contents of this book, while all were discussed at length both during the conference from which this book was derived and in the concluding workshop. A consensus and answers to these questions form the final chapter.

It is not likely that the reader of this book will find answers to all questions concerning waste problems. It is hoped, however, that some questions will be answered and that, as the reader improves his knowledge in this field, he will be able to make a more thorough evalution of the alternative possibilities, as well as deriving a better understanding of the basic premises of household waste management.

2 Development of Treatment and Disposal Methods of Solid Wastes in Europe in the 1980's

R. F. Millard

INTRODUCTION

To determine the changing pattern of waste disposal in the eighties it would seem logical first to attempt to forecast in which ways the quantities, densities and composition of domestic, commercial and industrial wastes will change. These, in turn, will be much influenced by the standard of living. Despite lower levels of economic growth in the major industrial countries, brought about largely by the greatly increased price of oil and other fossil fuels, and the diminishing availability of certain raw materials, living standards at the end of the seventies were far higher than at the beginning. It seems probable that in those countries with the highest standards the rate of future increases will continue to be relatively low, but new technologies will still enable improvements in standards to be made elsewhere in Europe.

At least 80% of domestic and commercial wastes are disposed of to land at the present time and for a large proportion of industrial waste there is no alternative to land disposal. In some countries crude dumping on marshes, in ravines or quarries is the practice. The rubbish is left to rot, and serves as a breeding place for flies, rats and other pests; is an eyesore, creates smell and possibly nuisance through smoke from fires.

In some less developed areas the crude dumping of domestic waste has a major redeeming feature — it enables the local people to exploit the vegetable fraction of the waste as a source of food for domestic animals, their pigs, goats, chickens, etc.

But as living standards increase so also does the demand for improved environmental benefits, and insanitary practices, including crude dumping of waste, become increasingly unacceptable to a more highly educated and well informed public.

A further factor that might reduce dumping is that as towns and cities expand and local waste dumps are filled the waste has to be transported to more distant sites at higher cost. When costs rise in this way the public will respond by demanding more efficient waste management. The service can no longer be left in the hands of a local entrepreneur. The local health authority will move in with statutory requirements and generally methods will gradually improve. The standard of waste disposal at the start of the eighties over a large part of Europe is, it is suggested, already to a publically acceptable environmental level and these standards can reasonably be expected to extend to the whole of Europe by the end of the decade.

Right up until the 1950's, some municipal authorities in Western Europe, including the UK, permitted the dumping of household waste in the seas around their coasts. Various international conventions, particularly those held in Oslo and London, have resulted in several countries banning disposal of waste into the sea, except under very special conditions. Slurries and sludges not containing toxic materials in unacceptable concentrations may be permitted, but most solid wastes must be transported far out to sea to deep water. Expense would rule out this practice for ordinary household and industrial wastes — it is limited therefore to particularly difficult special types of waste.

Dumping of waste into the Mediterranean still takes place by some countries but it is to be hoped that with the awareness of the increasing pollution of this ocean and following the 1980 United Nations Conference on the Mediterranean much more effective restrictions on this kind of dumping will take place.

What then is the future for disposing of Europe's waste for the next decade? It is suggested that the very high cost of fossil fuels makes the extraction of the combustible fraction of waste from domestic and commercial refuse and its conversion to a fuel an economic viability. Alternatively, the increasing cost of virgin wood pulp makes the extraction of the paper content for the production of reconstituted fibres an attractive proposition. And once plant and equipment has been installed to separate and classify waste it makes sense to recover the ferrous metal content for the production of new steel, etc., and possibly glass as cullet for use in the production of new glass.

TRANSPORT OF WASTE

The delivery of waste direct to the landfill site or treatment plant by the collection vehicle avoids the costs of double handling at a transfer station, but as available disposal sites become more distant from catchment areas so the case for intermediate transfer becomes unavoidable.

The requirement to use more distant and larger landfill sites, well remote from houses, is bound to continue to increase. The most usual method is to compact the waste into large-capacity road vehicles, but an alternative method finding favour for the more distant sites is to compact the waste into standard containers and load these on to special liner trains. Loading of the containers on to railway wagons takes place throughout the working day and at night an engine arrives at the waste transfer station to take the train load of containers to landfill — returning before the morning rush hour of commuters with a train of empty containers. Obviously the use of rail is limited to schemes where both the transfer station and landfill site are near to a rail link, but the increasing cost of diesel fuel, the restrictions on road vehicle drivers' working hours, together with the environmental benefit of keeping bulk transporters off residential roads, may make the use of rail for transporting waste more and more attractive.

Similarly, canals and rivers can be used for transporting waste from the centre of a large city — as is practised in, for instance, Central London. But when the waste is unloaded from the barge it must be dropped into a vehicle for transport to the tip face. This double or triple handling inevitably not only increases the costs involved, but creates litter from spillage and possibly pollution of the water course. The Rotterdam scheme whereby waste is packed into containers before transport by barge, in the Dutch case to an incineration plant, overcomes the environmental objections but at a still higher cost.

TREATMENT OF WASTE PRIOR TO DISPOSAL

The most simple and cheapest means of disposal is landfill of untreated wastes,

Jenkins Lane Road Transfer Station, (Courtesy — Greater London Council).

Transport Avenue Rail Transfer Station. (Courtesy — Greater London Council).

Loading barges at Cringle Dock River Transfer Station. (Courtesy — Greater London Council).

but because of escalating costs of transport more consideration will have to be given to treating the refuse before deposit. Pretreatment of waste inevitably adds to costs, but this may be offset by reduced transport costs due to increased pay loads, particularly when the length of haul to the disposal site is long. Added to which, there are some sites where planning authorities, and possibly water authorities also, would not tolerate the deposit of waste without some pretreatment.

Various processes have been tried, but the most successful at present, and likely to remain so, are:

shredding
baling
composting
incineration.

Shredded waste can be more easily compacted in a bulk vehicle than untreated waste, but a disadvantage compared to untreated waste is that if compacted into a standard container the load may not discharge so readily, especially at low temperatures.

Waste treated in this way is more easily controlled in landfill than untreated waste, is a lesser fire risk, less attractive to rodents, insects and birds, and less offensive visually during the landfill operations. Some sites, which would otherwise be unavailable for landfill on planning grounds, can therefore be brought in service and reclaimed. Shredding is of particular significance when the land is required for agriculture or recreational use after reclaiming without undue delay, as settlement is likely to be less pronounced and more uniform.

In high-density baling, untreated waste is compressed in a hydraulically operated machine into bales each weighing about one tonne. Self-sustaining bales can be formed, but a certain amount of gradual spingback occurs when the pressure is released. Nearly all takes place within the first 48 hours, normally without affecting the stability of the bales. Shredded waste can be formed into stable bales of similar density at lower pressures; they need to be banded with baling wires but expand less than self-sustaining bales.

Baling of waste in association with landfilling appears to be attractive. The bales can be deposited in a stacking process of blocks, so that natural expansion fills the intervening spaces, thereby forming a firm surface to the site, and maximum use of the space is achieved. It seems likely that shredding and baling will play a larger part in waste disposal practice in the eighties than in the past.

Only a small proportion of household and commercial waste is composted. The restricting factors include the cost of screening out unsuitable constituents (glass, plastics, metals, etc.) and pulverizing the remainder into a fine particle size. The percentage of rejects is high, and, in addition, long-term markets for the end product are difficult to achieve and sustain.

Compost from town refuse must be stored to mature for several months to allow inhibiting agents to leach out; it is then useful as covering material on landfill sites or for the improvement of marginal land, but a check must be kept to ensure there is not build-up of heavy metals. Mechanical systems for composting of household waste, sometimes with market waste or sewage sludge, have been installed in several places throughout the world, and some have proved quite successful.

Composting may well continue to play an important part in waste disposal in agricultural areas and where soil is lacking in humus. However, urban waste usually mixed with commercial and some industrial wastes is seldom of much benefit as compost.

Burning of waste, either in its delivered form or after shredding, seems to have enjoyed a form of cyclical popularity. In the thirties it had its advocates, and it enjoyed another spell of popularity in the early seventies. Of more recent years the growing awareness of the environmental problems associated with burning crude waste, especially where the vegetable content is fairly high, coupled with the very high unit costs involved has led to a marked decline in the number of plants under construction.

Incineration of combustible waste sterilizes the residues and reduces the bulk. The volume is reduced by up to 90% and the weight reduced by approximately 60%. Municipal waste does, however, contain constituents which when burnt produce gases such as hydrogen chloride, which can create strongly corrosive acids. These problems could be overcome, at least in part, by the installation of gas scrubbing plant. However, this adds to the already considerable capital and operating costs of incineration.

Similarly, the installation of electrostatic precipitators to extract the dust and grit from the gases arising from incineration of waste is very costly, and although they are very efficient, especially when new, some polluting dust and grit is, inevitably, emitted to the atmosphere.

There are several ways in which heat can be recovered.

(a) *District heating.* Some incineration plants have been designed so that the heat may be used either to supplement a conventional boiler plant or as the principal source of supply. However, it is difficult to achieve a design which meets the conflicting needs of reliability, peak period demand and financial considerations.

(b) *The use of superheated steam to drive turbines to produce electricity.* Since the efficiency of converting waste to power is low, it is doubtful if the price paid for producing the power will cover its cost, and then only in very large plants operating 24 hours a day, 7 days a week.

(c) *Heat released from the cooling of incineration gases could be used for the drying of sewage sludge.* This might be advantageous if the dried sludge is subsequently to be incinerated, but there has been only limited research in this direction. Alternatively, the waste heat or steam can be used within the plant.

The increasing cost of fossil fuels makes the recovery of heat from burning rubbish more attractive, but it seems more likely that the route for the future will be to recover the combustible fraction of waste from the refuse and use this after treatment as a supplement to coal, as discussed later.

Where landfill space for untreated, shredded or baled waste is at a premium, there is a case for total incineration, but space must still be found for the residuals and incombustibles. It is not probable, therefore, that there will be any marked increase in incineration in the near future.

DISPOSAL ON LAND

Once it is accepted that landfill of waste, on suitable sites, can be carried out to acceptable standards, the decision to be taken is what degree of treatment, if any, is necessary before final deposit.

However desirable recycling of waste materials may be from an ecological point of view, it is an incontrovertible fact that not more than a very small fraction of domestic and commercial waste will be re-used or reprocessed in the next decade. However, recycling of certain types of industrial waste, uncontaminated with mixed refuse, is much more likely to increase. Most of the residual waste requiring to be disposed of will be used to recover what would otherwise be derelict land, marshes, etc.

There are a few fundamental factors to be considered in assessing potential landfill schemes:

(a) effect on local residents, wildlife and amenities of the area,
(b) possibility of water pollution,
(c) problems of gas generation,
(d) planned after-use of the site.

The first factor is assuming ever-increasing importance, especially from national and local amenity and environmental groups — to such an extent that some responsible people consider that the pendulum has swung too far towards protection. Certainly the days have gone when the enterprising owner of a worked-out quarry could contract to receive wastes irrespective of the nuisances likely to arise.

Considerable research was undertaken in the seventies on the behaviour of wastes, particularly industrial wastes, deposited in landfill sites. The associated risk of polluting water courses has been assessed, and scientific data have been obtained on which to base guidelines and form criteria for the selection, operation and after-use of sites.

Basically the research indicates that providing there is an unsaturated zone of sufficient depth beneath the landfill, without significant fissures giving direct hydraulic continuity to an aquifer, the degree of pollution should not reach unacceptable limits.

This research will probably allow responsible water authorities to agree to the use of more sites for landfill in the eighties than hitherto. Another factor influencing the use of sites is the introduction of liners to sites. Liners can be of natural materials such as clay or puddled chalk, or membranes of plastic. Decomposing waste will produce leachate: all sites must now be designed either to permit the leachate to be dispersed and diluted, or with liners, to be concentrated and contained and the leachate recirculated or treated.

A problem that is becoming more acute is that of gas generation. Until a few years ago, waste in the better-run site was tipped and spread by bulldozers or similar plant, and covered with a thin layer of inert material. The tendency now is for larger sites, and with this has come the much heavier steel-wheeled compactors. To obtain maximum use of the available space, the more recent landfill technique is to spread the waste in thin layers, some as little as 0.6 m deep, and to compress it by the heavy steel wheels of the compactor. The intermediate and final cover is also compacted to a high density. As the waste decomposes, gases form. In the past these have vented harmlessly to the atmosphere via the comparatively loosely packed waste and cover. The greater degree of compaction in modern sites prevents this happening, forcing gases to migrate laterally and allowing for potentially dangerous pockets of gas to accumulate. To prevent these effects, special features, such as vertical artificial vents filled with gravel, stone or some similar material, must be designed into landfill sites and these should be constructed at intervals across the site. It would also seem essential in future from the outset to construct a loosely packed trench around the perimeter of the site to prevent lateral movement of the gases away from the site.

These gases cause problems from the nuisance created by smell and the danger of fire and explosion in the pockets of methane. However, the rapid increase in the price of natural gas means that in the next decade some landfill

Static compactor with ISO container attached (Courtesy — Greater London Council).

sites might be purposely designed to promote anaerobic decomposition of the waste to maximise methane production for commercial exploitation.

In future, much more consideration will have to be given at the design stage to the planned after-use of the site. Pressure groups are powerful in several geographical areas and their influence is likely to spread. A community would more probably tolerate a landfill site if they knew for sure that it was to become a park or recreation facility within a stipulated period.

A point to take into account, however, is that ordinary household and commercial wastes can take over 30 years to decompose and any buildings such as sports dressing rooms and toilets need adequate ground level ventilation.

RECLAMATION AND RE-USE OF WASTE

Large quantities of factory waste are reclaimed for re-use in-house, and in other cases nationally organized waste exchanges have been instrumental in bringing together a producer of 'waste' and potential re-user.

Factory waste is usually of a fairly consistent nature and there is a regular output. Household and commercial waste, on the other hand, varies in composition with the season of the year, the geography of the area, the type of area and many other factors. On average, however, it will often include, by weight:

paper and board	20-40%
plastics	5%
ferrous metals	10%
glass	10%
vegetable matter	20-40%

The balance will be textiles, dust and general mixed rubbish. The moisture content of household waste varies with the weather and season; it is usually between 15% and 30%.

Some collection authorities find it profitable to collect waste paper simultaneously with domestic waste, in separate trailers towed behind the vehicle, but few collect anything other than paper in this way. Other authorities arrange special collections of paper and cardboard, in separate vehicles, from shops, offices and factories, where the supply justifies it, but most paper recovered from these sources is by private contractors. The problem faced by the authorities on separate collections at source is simply one of economics — it is far from easy to collect sufficient paper from a given area to make a profit. Voluntary organizations play an important part in paper recovery when demand from the board-making industry is high.

Once the waste has been mixed in the dustbin or disposable sack, its separation by mechanical means is difficult. There is a limited demand for cans recovered from the waste; at present they command a low price, even when baled, but this may well rise as the industry picks up again.

The demand for glass is limited. Considerable research is under way to find new uses for mixed broken glass, but the raw materials are cheap and plentiful so that economic recovery is difficult. Some collection agencies arrange for containers to be placed in depots and car parks, etc. The public are invited to place their bottles in these containers, keeping the coloured glass separate from the clear.

The paper that can be recovered mechanically from municipal waste is often contaminated with food and mixed with plastics and textiles. Its value to board manufacturers as a raw material is therefore limited.

However, this part of waste, together with the plastic and textiles, is over 30% by weight of the total input, has a high calorific value and may be used as a supplement to conventional fossil fuels in existing boilers. To be marketable, this waste-derived fuel must have qualities acceptable to the user. Oil, gas, electricity and coal are well defined fuels, but fuels produced from waste are physically and chemically different from their fossil fuel counterparts, and characteristics such as ash content, heat value, corrosiveness and storage properties have to be acceptable to the user.

Assuming that up to one third by weight of municipal waste could be extracted for use as a fuel and sold at an economical price, the principal benefit to a waste disposal authority would be the reduced quantity of residue, especially by volume, that would need to be disposed of and the consequent overall saving in final disposal cost. The main benefit to the industrial user would be a reduction in fuel costs.

Both private and public sectors are showing considerable interest in the production of a waste-derived fuel, either in a shredded and fluffed form or as pellets. The eighties will show whether the production of waste-derived fuel in commercial quantities can be economically viable.

CONCLUSIONS

Analyses of waste throughout the world indicate that as the standard of living rises so does the volume of wastes produced. And as industry develops, so does the complexity and volume of wastes it produces.

Disposal authorities throughout the world have in the past, and will no doubt in the future, seek a panacea to solve the problem of getting rid of its unwanted wastes. Various techniques have been tried.

Some limited research into pyrolysis of municipal waste has been carried out, mainly in Japan and the USA, and to a more limited extent in Germany and the UK. It seems unlikely that pyrolysis will play more than a very marginal part in domestic waste treatment in the next decade, but its future for certain types of industrial waste, such as car tyres, seems more promising. Similarly, the technique of fluidized bed incineration seems likely to be developed, but with more concentration on special industrial wastes than on municipal waste.

What does appear to be certain is that by far the largest proportion of domestic, commercial and industrial wastes — probably over 80% by weight — will continue to be disposed of to land. There should, and probably will, be greater emphasis on the standards employed, the degree of pretreatment and the planned after-use of the sites involved.

The demand for the use of the combustible fraction of waste — the paper, plastics and textiles — may well result in a proliferation of resource-recovery plants, for when part of the waste can be sold as a fuel it makes economic sense to recover the ferrous metals and possibly also other constituents.

Improved technology and public concern together may also lead to the construction of plants to recover as a primary function the paper fraction for further treatment with the aim of producing fibres for new board and paper manufacture.

There is at present no universal solution to our waste disposal problems. There is not likely to be one by 1990.

Mr. Millard is General Manager in charge of the Solid Waste Management service of the Greater London Council. He is responsible for the transfer of some 11 000 tonnes of waste each day from collection vehicles to road bulkers, river barges and liner trains and for its treatment by shredding, incineration or landfill. He was previously superintending engineer at the Department of the Environment and represented the UK on the research sub-committee on waste in the EEC at Brussels.

3 The Importance of Packaging in Household Waste

Eveline Hajek

INTRODUCTION

There is a constantly growing problem of household waste disposal in most countries of the civilized world. The increasing consumption of packaging materials is blamed for this development — not unjustly, as will be shown later.

To give a view of the dimensions of the problem thrown-away packages cause on household waste, this chapter will report on the packaging share of household waste and its composition. On the basis of these data, something can be said about the future development of this packaging share. Finally, the question will be discussed of how to reduce the packaging share of household waste in the future — if not absolutely, at least relatively — and what action should be taken to achieve this aim.

THE COMPOSITION OF HOUSEHOLD WASTE IN AUSTRIA

The composition of Austrian household waste was thoroughly investigated in 1973. The aim of this investigation was to find out empirically the packaging share of household waste[1].

The groups of investigated materials shown in Table 3.1 were fixed, the components of the household waste allocated to these groups, and their weight

TABLE 3.1 Analysis of Household Waste – Groups of Investigation

Packaging waste	Remaining household waste
paper	vegetable refuse
board, corrugated board	textiles, leather, rubber
nonferrous metals	paper
ferrous metals	board, wood
glass	mineral components
plastic films	metals
plastic containers	plastics
aerosol bottles and cans	
wood, cork	
others (e.g. textiles)	

and volume evaluated. As a main result of this refuse analysis, it was shown that the packaging share of household waste was 32% by weight and 52.2% by volume. These figures confirm some other authorities, who estimated the packaging share at about 30%. Similar figures are known from the USA[2], where the packaging share of wastage varied during the years 1971 and 1975 between 32.5% and 34.8% (Table 3.2). In the Federal Republic of Germany, the packaging share of wastage is between 30% and 34%[3] and in Switzerland packaging share of wastage is estimated at about 28%[4].

TABLE 3.2 Packaging Share of American Household Waste

	1971	1972	1973	1974	1975
Packaging share of household waste (% by weight)	33.4	34.6	34.8	33.7	32.5

The composition of the packaging share of household waste is shown in Table 3.3. Packages of glass and packages of paper and board are responsible for two thirds of the packaging waste by weight, while packages of paper and board, and especially plastic packages, are the main components of packaging waste by volume.

TABLE 3.3 Composition of Packaging Waste

Type of packaging material	Percentage by weight	Percentage by volume
glass	34.2	10.0
wood	1.6	1.8
plastics	15.5	31.7
metals	16.4	14.3
paper, board	31.8	41.6
textiles	0.5	0.6
Total	100.0	100.0

The importance of the packaging material groups for the total household waste is illustrated in Table 3.4. By weight, packages of glass and of paper and board hold the largest share, with 10% each, followed by metal and plastic packages with approximately 5% each and packages of wood and textiles with less than 1%. By volume, packages of paper and board compose 20% of household waste, plastic packages 16%, metal packages less than 10% — there are doubts that the metal share is as high, because when the waste is separated before incineration less metal is found — and wood and textile packages with minimal contributions.

Figures for household waste and packaging waste per inhabitant in Austria are shown in Table 3.5. All the glass waste of 25 kg per annum is packaging waste. 12 kg of the 14 kg plastic waste is caused by packages, 12 kg of the 20 kg metal waste is tin cans, foils and so on, and 25 kg of the 71 kg of paper is contributed by paper bags, cardboard boxes, etc. The considerable remainder of this material group is all the newspapers, periodicals, telephone books and folders.

TABLE 3.4 Shares of Packaging Materials of the Total Household Waste

Type of packaging material	Percentage by weight	Percentage by volume
glass	11.1	5.2
wood	0.6	1.1
plastics	4.9	16.5
metals	5.1	7.4
paper, board	10.4	21.7
textiles	0.1	0.3
share of packaging waste	32.2	52.2

TABLE 3.5 Household Waste and Packaging Waste per Inhabitant in Austria in 1975

Types of packaging material	Household waste (wet) kg/inhabitant/year	Packaging waste/kg/ inhabitant/year
glass	25	25
wood	5	not known
plastics	14	12
metals	20	12
mineral components	33	—
paper	71	25
textiles	19	not known
vegetable refuse	51	—

More recent figures can be gathered from an investigation carried out in the years 1977 to 1979[5]. In a limited Viennese community with 13 470 inhabitants and 4342 households, a separate collection of salvage was carried out and analysed. Because of the fact that specifically metal, glass and plastic packages were collected separately, only these shares of household waste are known. The figures obtained are shown in Table 3.5, compared with figures for 1973. An

TABLE 3.6 Packaging Share of Household Waste, Weight %

Type of packaging material	1973	1977/78
glass	11.1	8.3
plastics	4.9	8.8
paper, board	10.4	12.3

interesting fact is that compared with 1973 the share of plastic packages has increased considerably. On one hand this can be explained by the increased use of plastic for packaging purposes and on the other hand by the urban structure of the community studied.

Factors Affecting Quantity and Quality of Waste

In statements about the quantity and composition of household waste, the variability of composition must not be forgotten. Above all, quantity and quality of the household waste depend on:

(a) the size of the community — the waste quantity per inhabitant increases from 110 kg in communities with less than 1000 inhabitants to 247 kg in communities with more than 100 000 inhabitants[6],
(b) the changing way of living, the living standard and the influence of the general economic condition,
(c) the season,
(d) the differences in social structure (average age, education) and economic power of a region,
(e) the number of people who live in one household — one-person households cause two or three times the specific waste quantity per person than households with five people[7],
(f) the volume of the dustbins — if small dustbins of 25 l are used, it has been found that the quantity of waste per inhabitant per annum is less than if larger ones of 110 l are used[8]. The reason for this might be that housewives look for other ways to get rid of their waste and put it to use; or, for instance, they may not take home outer packages from the supermarket because they are aware that they can't be disposed of without being cut up.

Concerning the factors that influence packaging waste, we differentiate between

(a) factors influencing the volume of packaging waste, and
(b) facts influencing the weight of packaging waste.

The following relations can be stated:

The weight of packaging waste depends on
- (a) number of packages disposed of,
- (b) weight of packages disposed of,
- (c) size of packages disposed of.

The volume of packaging waste depends on
- (a) number of packages disposed of,
- (b) size of packages disposed of,
- (c) stability of packages disposed of.

The number of packages disposed of in the dustbins is again dependent on
- (a) production of goods by weight as a function of the development of the population and the general economic development,
- (b) ratio of unpacked to packed goods,
- (c) ratio of one-way bottles to multiple-way bottles,
- (d) number of re-used packages,
- (e) number of packages collected separately before they become household waste.

TRENDS IN THE QUANTITY AND SHARE OF PACKAGING IN HOUSEHOLD WASTE

To be able to say something about the future development of packages in household waste, it would be interesting to know how the packaging industry

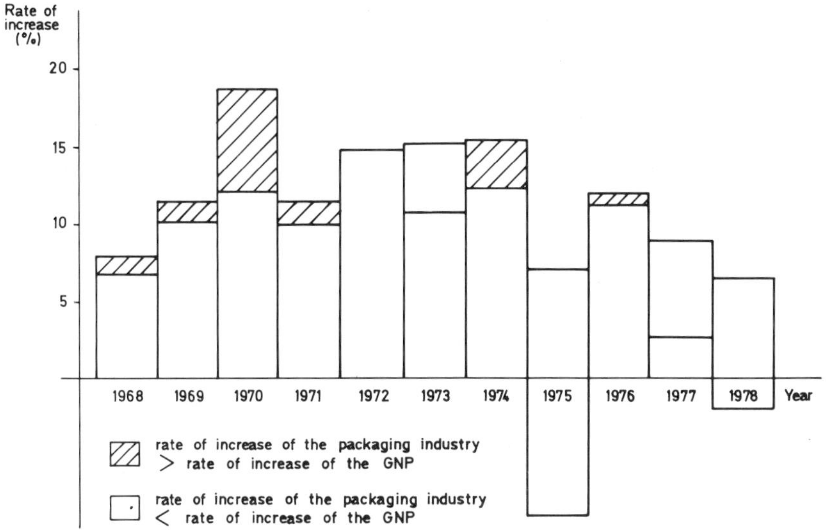

Fig. 3.1 Rates of increase of the packaging industry compared with the rates of increase of the GNP.

will continue to develop. The fast development of the packaging industry since 1950 is due to increasing consumer demand, the development of modern distribution systems and the establishment of self-service shops. The transportability of goods was facilitated essentially by prepacking; the storage of several products (e.g. vegetables) was thus possible and the self-service system is not feasible without offering goods in a prepacked form.

Therefore packaging undertakes not only the quantitative function — goods are offered in consumer-like quantities — and the qualitative function — goods are offered in an adequate quality — but also the information of the consumer. It is most probable that the past double-figure rates of increase of packaging production will no longer be achieved in the eighties. It may be that Europe is coming to (or has already reached) a saturation level. There are indications for such a saturation point:

(a) Since 1971 the packaging industry has increased slower than the GNP[9] (see Fig. 3.1). Exceptions were only the years 1974 and 1976. This was caused by the energy crisis, whereupon industrial firms with goods to be packed filled their warehouses with packaging material. Following this, in 1975 there was a negative purchase of packaging which was compensated for in 1976.
(b) The packaging production share of the GNP is decreasing constantly. The year 1974 is again an exception, for the same reason as mentioned above. (See Table 3.7). The same tendency is shown in the Federal Republic of Germany (see Table 3.8).
(c) The ratios of increase of the packaging production by weight are declining (see Fig. 3.2).

These developments disprove the assumption that packaging processes have become basically more sophisticated. Nevertheless, it can be assumed that the rates of increase of the packaging waste will correspond in future to the rates of increase of the goods produced in an economic system. This is due to the fact

TABLE 3.7 Share the Packaging Production of the GNP (Austria)

	1972	1973	1974	1975	1976	1977	1978
GNP (in milliard Sch)	473.45	532.29	610.08	652.64	722.60	785.67	835.00
Production of the packaging industry (in million Sch)	7.127 941	7.585 324	9.563 773	8.888 862	9.936 862	10.165 198	9.93 8251
Share (%)	1.51	1.43	1.47	1.40	1.38	1.29	1.19

TABLE 3.8 Share of the Packaging Production of the GNP (BRD)

Years	Share (%)
1957 – 1961	1.77
1962 – 1966	1.76
1967 – 1972	1.68
1973	1.55

that there are only a few groups of products that are partly or wholly unpacked. These product groups are primarily:

fresh fruits and vegetables
fresh meat, sausage and fish
baker's products

If these product groups are sold to a greater extent prepacked, the influence on the packaging waste will, however, be a limited one. The product groups in question have a quick turnover, but the packaging means used today — paper trays and foils — allows the assumption that more prepacked fresh food doesn't mean a considerable increase in packaging waste, either by weight or by volume. Also, although these products are not generally sold prepacked today, they are nonetheless packed in some way.

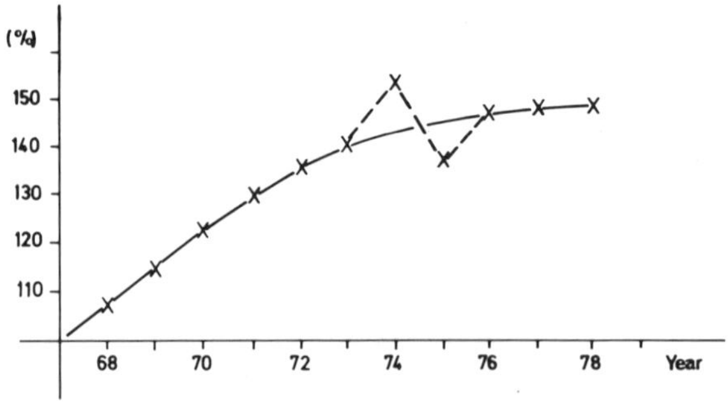

Fig. 3.2 Increase of packaging production by weight.

Socio-economic Changes

In trying to forecast the future consumption of packaging materials, findings must be based on the development of the product groups bought by private households, taking into particular consideration prepacked product groups for daily use. These product groups influence household waste much more than product groups that are bought only periodically (electrical appliances, household goods) and to a lesser extent (furniture). The above-described product groups are:

foods
cosmetics
clothes (partly)

To estimate the development of consumption of these product groups, the expected expenses for consumption and the development of the population can be used as indicators.

As in the last 20 years, the share of the total expenses per inhabitant produced by the product groups relevant for packaging will continue to decline[10]. Between 1954 and 1959, 44.7% of expenses for consumption were spent on food and beverages; in 1985 this figure will shrink to 22.3%. The development of expenses for clothing and cosmetics will remain constant. There will be increases of productivity in the education, traffic and household goods sectors. The packaging-intensive consumption areas will decline — at least relatively — in importance. (See Table 3.9 and Fig. 3.3).

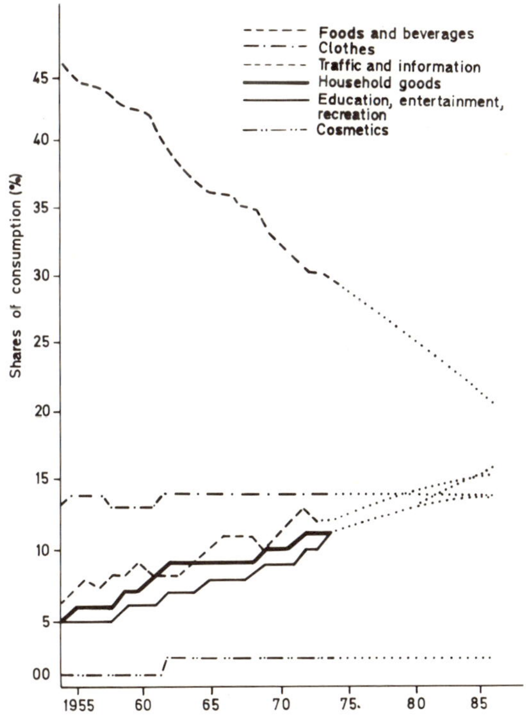

Fig. 3.3

TABLE 3.9 Development of the Structure of Consumption (Austria)

	1974		1980		1985	
	Sch/ inhabitant prices 1964	%	Sch/ inhabitant prices 1964	%	Sch/ inhabitant prices 1964	%
food and beverages	8335.7	29.9	9015.9	26.2	9292.6	22.3
clothes	3826.3	13.7	4759.4	13.8	5734.3	13.7
household equipment	3200.3	11.5	4548.2	13.2	6169.4	14.8
traffic	3299.6	11.8	4634.9	13.4	6360.6	15.2
education, entertainment and recreation	3196.7	11.5	4418.2	12.8	5712.5	13.7
cosmetics	543.1	2.0	789.6	2.3	1054.8	2.5
other goods and services	5449.3	19.6	6309.4	18.3	7423.1	17.8

The absolute development of this part of consumption depends on the development of the population. As the population of Austria has increased constantly up to the middle of the 1970's, no further increase is expected till the year 2000. Even a small decline in the population would be no surprise[11]. (Table 3.10). The declining population and the increasing private consumption per inhabitant mean that the share of the total private consumption relevant for packaging will not cause an undue waste problem.

TABLE 3.10 Population Development 1951–2001

	1951	1961	1971	1981	1991	2001
Number of inhabitants	6.933 905	7.073 807	7.456 403	7.445 200	7.356 900	7.216 800
change (%)		+0.2	+5.4	−0.2	−0.2	−0.2

A further important factor of influence is the size of the household. The change of the average size of households extrapolated to the end of the century shows a continuing trend to smaller households (see Table 3.11). In 1951 the average household size in Austria was 3.1 people; in the year 2001 there will be only 2.6 people per household.

TABLE 3.11 Private Households 1951–2001

Type of household	1951	1961	1971	1981	1991	2001
One-person household	386.0	453.5	623.0	635.7	623.6	609.2
Household with more than one person	1819.2	1852.3	1912.9	1968.0	2097.6	2190.1
Total number of private households	2205.2	2305.8	2535.9	2603.7	2721.2	2799.3
Average size of household	3.11	3.02	2.90	2.82	2.72	2.63

The change in the number of people per private household and in its structure, corresponding to smaller consumption units, lead to smaller packages and therefore to an — at least relative — increase in packaging waste. The tendency of more and more housewives to go out to work and the restricted storage possibilities in the household lead to convenience packages, a reduction in size of single purchases and therefore also to an increase in packaging waste.

Another influence on household waste is the change of shopping habits. The trend to do the shopping once a week at the weekend increases. This means larger packaging units.

The tendency to larger packaging units meets the intentions of the packaging industry; because of this, the packaging cost per quantity unit can be reduced. This is especially important where cheap products such as mineral water, fruits and biscuits are concerned.

Techno-economic Changes

Industrial firms also show a trend to reduce the weight of packages; for example, the reduction of the weight of glass containers and cans — in 1960, a glass bottle had an average weight of 507 g, and in 1972 its weight was only 403 g (Table 3.12). The weight of tin cans has declined between 1951 and 1979 from 83 g to 32 g; this corresponds to a reduction in weight of more than 60%[13] (Table 3.13).

A relative reduction of packaging waste has taken place and will continue.

TABLE 3.12 Average Weight of Glass Containers[12]

Year	Weight (g)	Change (%)
1960	511	
1961	507	−0.8
1962	499	−1.6
1963	505	+1.2
1964	506	+0.2
1965	509	+0.6
1966	484	−4.9
1967	458	−5.4
1968	432	−5.7
1969	412	−4.6
1970	415	+0.7
1971	405	−2.4
1972	403	−0.5
Total weight reduction since	1960	−21

Lighter-weight and better-stackable packages also meet the intentions of the filling industry. Examples from the past are the complete substitution of milk cartons for multiple-way glass bottles and the substitution of corrugated board boxes for wooden boxes. Wherever it is possible to pack goods in this manner,

TABLE 3.13 Development of Can Weights (0.35 l) 1951 – 1975

Year	Can type	Weight (g)	Percent of 1951
1951	three-piece tin can with seamed lid	83.0	100
1965	three-piece tin can with pull-tab lid	70.1	84.5
1967	three-piece tin can with pull-tab lid	63.6	76.6
1969	three-piece tin can with pull-tab lid	62.7	75.5
1972/75	three-piece drawn tin can with pull-tab lid	48.4	58.3
1975/76	two-piece tin can with pull-tab lid	38.0	45.8
1976	two-piece tin can with pull-tab lid lightweight, US-type	34.5	41.6
1979	lightweight tin can with pull-tab lid (in development)	32.0	38.3

the tendency will continue, and for this reason packaging waste will decline by weight.

To summarise briefly the development of the essential factors influencing packaging waste:

(a) the population figure is stagnating,
(b) production of consumer goods relevant for packaging waste will increase slower than production in other sectors of consumption,
(c) apart from a few exceptions, products for consumption are already entirely prepacked,
(d) the frequency of shopping is declining,
(e) the storage possibilities in the household are becoming smaller,
(f) the size of the average household is decreasing and the number of small households increasing,
(g) the number of working housewives is increasing.

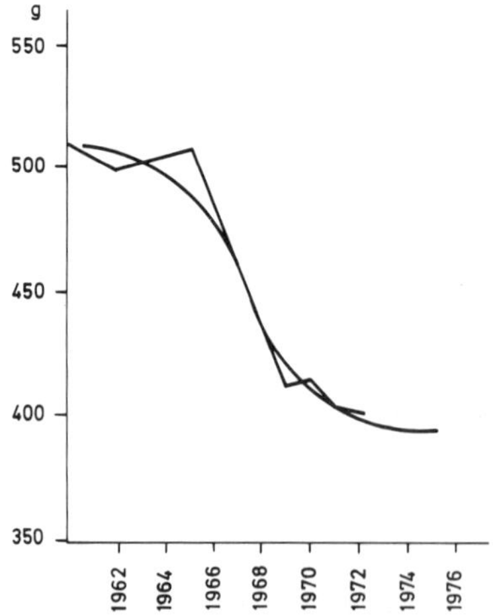

Fig. 3.4

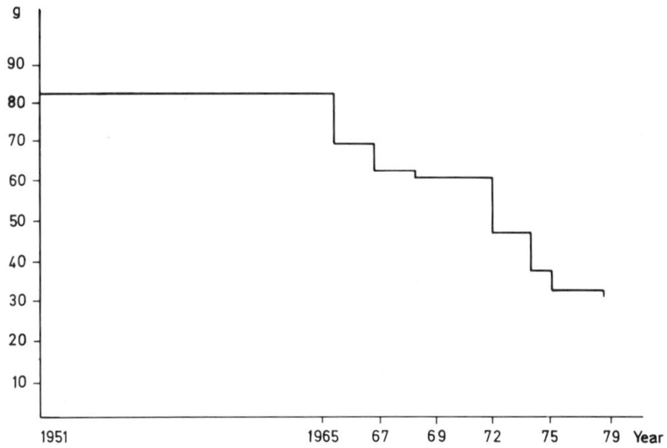

Fig. 3.5

These developments allow the assumption that the 'packaging waste mountain' has most probably achieved its climax. Thus it is hopeful that the ratios of increase by weight will be more moderate than in the past.

POSSIBILITIES FOR REDUCING PACKAGING WASTE

Firms producing goods to be packed or filled decide which sort of package has to be used and therefore have an influence on the disposal problem. Above all, their decision depends on the costs of the different packages. As the reserves of rationalisation in the production area are largely exhausted, attention has in the recent past been drawn to the distribution area and, with this, to packaging. In former times goods were packed 'anyhow'; the purpose was to provide that the goods reached the consumer undamaged. Nowadays, firms are trying to make the packaging process optimal, with the aim that the package should be sufficient and also as cheap as possible. This means that firms are not only considering the functions a package must fulfil, but also the costs; thus the packaging process between the producer of the package and the consumer of the packaged good will be optimal in the future. But this means also that the fact that the package has to be disposed of after use is not taken into consideration. The costs packages cause when they have become waste need not be born by producers of packages, filling firms and trade.

The packaging industry and the filling industry can be stimulated to change their packaging habits only by the following events:

(a) *If the consumer begins to include the sort of package used in his buying decision.* Today such behaviour can be expected only as far as his convenience is concerned. For instance, beer in cans is bought to be drunk at sea or at the tennis court, while for consumption at home beer is stored in beer crates and returnable glass bottles. Motives such as material saving, energy saving and waste avoidance are not yet such a part of rationalised thinking as to influence the buying decision. Therefore it cannot be expected that consumers will influence the industry to use packages that can be disposed of without problems and with minimal costs.

(b) *If prices on the raw material market increase.* Increasing prices, which indicate a raw material shortage, can support efforts to reduce the weight of packages, as mentioned above. They can also make it worthwhile to use more salvage materials, provided that increasing demand encourages recycling and the costs of collection and processing are lower than the prices of new raw materials.

(c) *If governments take action.* The aim to reduce packaging waste can be achieved by:

   (a) substitution of packages by others that affect the packaging waste less than previous ones,
   (b) increase of trippage of multiple-way bottles,
   (c) technical progress,
   (d) recycling.

Table 3.14 shows that a reduction of packaging waste by weight can be achieved by:

(a) the use of packages that are lighter in weight, both one-way and multiple-way packages,

(b) the use of more multiple-way packages, if a high trippage rate can be expected,

TABLE 3.14 Secondary Aims to Reduce Packaging Waste and their Efficiency

OW-P = one-way package; MW-P = multiple-way package; T = trippage.

Substitution of MW-P by other MW-P	Substitution of OW-P by other OW-P	Substitution of MW-P by OW-P	Substitution of OW-P by MW-P
Influencing the volume of packaging waste			
no influence	if flexible packages are substituted for solid packages, i.e. cans are substituted by flexible containers	if flexible packages are substituted for solid packages, i.e. multiple way bottles are substituted by flexible containers	in any case if MW-P are substituted for OW-P of the same material if MW-P are substituted for OW-P of a different material, the achievement of the aim depends on the trippage rate
Influencing the weight of packaging waste			
if $\frac{g\ \text{MW-P}_1}{T_1} < \frac{g\ \text{MW-P}_2}{T_2}$ i.e. if MW-plastic bottles are substituted for MW-glass bottles	if $g\ \text{OW-P}_1 < g\ \text{OW-P}_2$ i.e. if cans are substituted for OW-glass bottles	if $g\ \text{OW-P} < \frac{g\ \text{MW-P}}{T}$ i.e. if trippage rate is small and OW-P are light in weight	in any case if MW-glass (plastic) bottles are substituted for OW-glass (plastic) bottles, i.e. if plastic boxes are substituted for wood boxes

(c) the increase of trippage rate of multiple-way packages,
(d) the separate collection of salvage in households.

A reduction of packaging waste by volume can be achieved by

(a) the use of flexible packages, if possible,
(b) the use of more multiple-way packages, if a high trippage rate can be expected,
(c) the increase of trippage rate of multiple-way packages,
(d) the separate collection of salvage in households.

Which actions[14] achieve these aims? Only the following aims are of interest:

(a) substitution of one-way packages by multiple-way packages,
(b) increase of trippage rate,
(c) technical progress,
(d) recycling.

The other aims are not interesting for the following reasons.

(a) Substitution of multiple-way packages (mainly glass bottles) by other multiple-way packages (i.e. plastic bottles) is not presently technically mature. Multiple-way plastic bottles for milk are currently under test in the BRD.
(b) Substitution of one-way packages by other one-way packages can hardly be influenced by the government; additionally, the filling industry uses for their own interest packages that are as lightweight as possible.
(c) Substitution of multiple-way packages with a low trippage rate by one-way packages has already been perfected.

Increase of trippage of MW-P	Technical progress	Recycling
in any case	no influence	if used and empty packages are collected separately in the household
in any case	if technical progress leads to a reduction in weight	if used and empty packages are collected separately in the household

Mechanisms for Control

Actions of the government which could aim at a substitution of one-way packages by multiple-way packages are

bans
restrictions
taxation.

Restrictions are the necessary requirement of retaining, reducing and limiting the rates of increase. A ban and the three methods of restriction can influence the following sectors:

production of one-way packages
filling into one-way packages
trade with one-way packages (filled)
import of one-way packages (filled or empty)

This results in 17 methods, which are shown in Fig. 3.6. An increase of the trippage rate can be achieved by the following actions:

unification of deposits
standardization of packages
obligation to mark one-way and multiple-way bottles
consumer information
the collecting of deposits

Technical progress like the promotion of the development of material-saving technologies are also able to reduce packaging waste.

Finally, material recycling promoted by consumers information and the collection of deposits of one-way packages as well as the separate collection of salvage are stated as actions to reduce packaging waste.

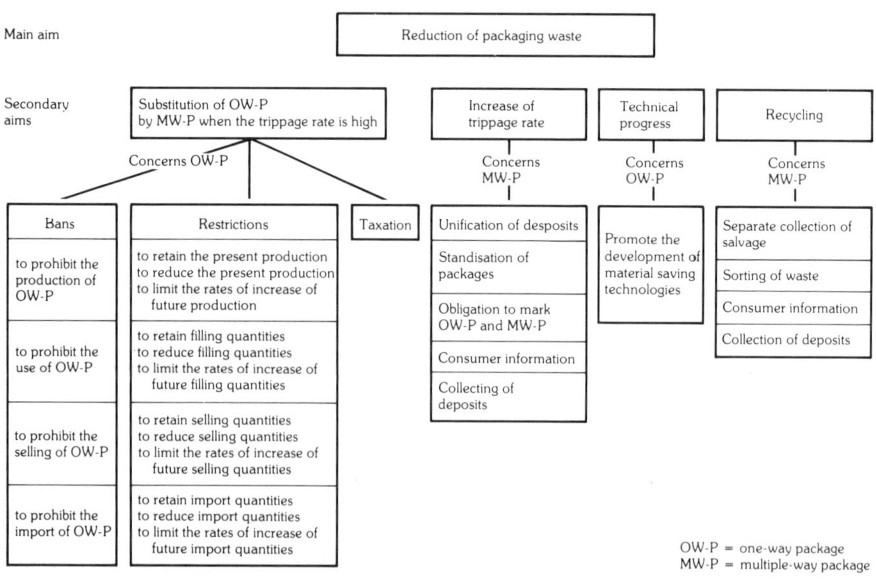

Fig. 3.6 Action to achieve reduction of packaging waste.

Efficiency of Control Mechanisms

Are the actions mentioned above efficient, and if so under what conditions?

The efficiency of bans and restrictions on one-way packages depends on several factors:

(a) *The sort of restriction chosen.* Bans represent the most effective intervention on the market, the limitation of further rates of increase the less effective one. This leads to quantitative differences in the waste reduction rate.

(b) *The development of the market.* It is possible to differentiate between a constant, an increasing and a decreasing market (see Table 3.15).

TABLE 3.15 Efficiency of Action at Different Market Situations

Assumptions: the substitution of one-way packages by multiple-way packages achieves the purpose; without market intervention the ratio of one-way packages to multiple-way packages remains constant.

Action	Market		
	Constant	Increasing	Decreasing
bans	efficient*	efficient*	efficient*
reduction	efficient*	efficient*	efficient°
retaining	inefficient	efficient*	inefficient
limitation of ratio of increase	inefficient	efficient°	inefficient

*interesting action in the respective market situation.
°interesting action in the respective market situation only under certain conditions.

Bans of one-way packages are efficient in all three kinds of market situations.

A reduction of one-way packages achieves the aim only in a constant or an increasing market. In a decreasing market the purpose can be aimed at only if the reduction rate of one-way packages allowed to be sold is larger than the shrinking rate of the market.

Retaining the number of one-way packages achieves this only in an increasing market; limitation of the ratios of increase does so only if the rate of increase of the one-way packaging market is larger than the allowed extention of the number of one-way packages to be put onto the market.

With declining intensity of action, the quantitative effectiveness is decreased and the dependence of the efficiency on the respective market situation increases.

(c) *The branch of economy, the action to be applied.* The branches of economy where the actions in question can be applied are

   (a) the packaging industry (production of one-way packages can be restricted)
   (b) the filling industry (filling of e.g. beverages in one-way packages can be restricted)

(c) the trade (selling of certain products in one-way packages can be restricted)
(d) the importer (importing of empty one-way packages or certain products in one way packages can be restricted).

The closer to the consumer in the hierarchy of the economic system that an action is applied, the more efficient the control of the number of one-way packages placed on the market. Action applied to the packaging industry can be circumvented by importing empty one-way packages or filled one-way packages. Only action applied to the trading sector enables a strict control of the number of one-way packages placed on the market. Such action can not be circumvented.

(d) *The reaction of the affected firms.* The possible reactions of an affected firm can be explained by an example where the filling of carbonated beverages into one-way bottles is prohibited with the aim of reducing packaging waste. If imported products filled in one-way packages are substituted for one-way packages of the same product, which can be no longer filled in the country, the same number of one-way packages is disposed of and the action is inefficient. If products filled in multiple-way packages are used instead of the forbidden one-way packages, the action is efficient if the trippage rate is high enough. If there is no substitution of the forbidden one-way packages and the product disappears from the market, the action is efficient in any case.

Whether a reduction of packaging waste can be achieved by taxation of one-way packages or of one-way packages and multiple-way packages depends first of all on the reaction of the consumer; his willingness to bear the additional costs caused by the amount of tax. If he is not willing to do this, in the long term one-way packages will be substituted by multiple-way packages.

Regarding the efficiency of the remaining actions,

collection of deposits on multiple-way packages
unification of deposits,
standardisation of packages,
separate collection of salvage,
consumer information,

more definite statements can be made.

There are no complicated conditions under which the action is efficient, but one must take into consideration the reactions of the consumer. Provided that the consumer accepts the aim that the environment should be protected, these actions will lead in any case to a reduction of packaging waste.

Tables 3.16 and 3.17 summarise once again the discussed actions and their efficiency.

To sum up, it can be said that the most purposeful actions to reduce packaging waste in households are the promotion of trippage rate, of materials-saving technologies and of the separate collection of salvage in households. These actions achieve the purpose in any case. Bans, restrictions and taxation are efficient only under certain conditions; we must also take into account that this would mean costs caused by the execution and control of such a law and would also be a considerable intervention of the market. Finally, it should be pointed out that there are a number of possibilities for management of the waste problem. Whether waste avoidance or the separate collection of packagings from household waste should be promoted, or whether it is more reasonable to recycle raw materials by

BLE 3.16 Efficiency of Actions that Aim for Substitution of One-Way Packages by Multiple-Way Packages

tion	Possible substitutions	Purpose achievement		
		Reduction of waste[1]	Savings in raw material[2]	Savings in energy[3]
probit production	OW-P* inland → OW-P* foreign	i	e	e
retain production	OW-P* inland → MW-P* inland	(e)	(e)	(e)
reduce production	OW-P° (P* inland) → OW-P° foreign	i	e	e
limit the rates of increase of production	no substitution[4]	e	e	e
	combinations[5]			
prohibit use	OW-P° (P* inland) → OW-P° foreign	i	e	e
retain filling	OW-P° (P* foreign) → OW-P° foreign	i	i	i
reduce filling	OW-P° (P* inland) → MW-P° inland	(e)	(e)	(e)
limit the rates of increase of filling	OW-P° (P* foreign) → MW-P° inland	(e)	i	i
	no substitution[4]	e	e	e
	combination[5]			
prohibit the selling	OW-P° (P* inland) → MW-P° (P* inland)	(e)	(e)	(e)
retain selling	OW-P° (P* foreign) → MW-P° (P* inland)	(e)	i	i
reduce selling	OW-P° foreign → MW-P° (P* inland)	(e)	i	i
limit the rates of crease of selling	no substitution[4]	e	e	e
	combination[5]			
prohibit imports	OW-P* foreign → OW-P* inland	i	i	i
retain imports	OW-P* foreign → MW-P* inland	(e)	i	i
reduce imports	OW-P° foreign → OW-P° (P* inland)	i	i	i
limit the rates of increase of imports	OW-P° foreign → MW-P° (P* inland)	(e)	i	i
	no substitution[4]	e	e	e
	combinations[5]			
xation	no substitution by MW-P°	i	i	i
	OW-P° foreign → MW-P° (inland)	(e)	i	i
	OW-P° (P* foreign) → MW-P° (inland)	(e)	i	i
	OW-P° (P* inland) → MW-P° (inland)	(e)	(e)	(e)

= efficient i = inefficient
roduce of one or more packaging materials.
esult of the combination of goods to be packed and the packaging.
e) = efficiency depends on the trippage rate.

[2](e) = different raw materials can not be compared.
[3](e) = efficiency depends on the trippage rate and the packaging-dependent energy consumption at warehousing, transport and trade.
[4]This means that the product in question disappears from the market.
[5]If different substitutions take place, the efficiency of the action can be determined for the combination.

TABLE 3.17 Efficiency of the Remaining Actions

Action	Objective	Reduction of waste	Savings in raw materials	Savings in energy
collecting of deposits (MW-P)	increase of trippage rate	e	e	e
unification of deposits	increase of trippage rate	e	e	e
standardisation of packages	increase of trippage rate	e	e	e
obligation to mark OW-P and MW-P	increase of trippage rate	e	e	e
consumer information	increase of trippage rate	e	e	e
promote the development of material saving technologies	technical progress	e	e	e
promote the development of energy saving technologies	technical progress	i	i	e
separate collection of salvage	recycling	e	e	e
sorting of waste	recycling	i	e	e
the collecting of deposits (OW-P)	recycling	e	e	e
consumer information (when separate collection of salvage)	recycling	e	e	e

e = efficient   i = inefficient

separation in special plants or even to incinerate the whole household waste will be in the last analysis a question of economy.

REFERENCES

1. 'Verpackungsanteil am österreichischen Hausmüll, Untersuchung der quantitativen Zusammensetzung des österreichischen Hausmülls unter besonderer Berücksichtigung der Anteile der Verpackungsstoffe', *Schriftenreihe Verpackungsforschung*, Heft 2, Vienna 1974.
2. DARNAY, A. and FRANKLIN, W., *Role of Packaging in Solid Waste Management 1966 to 1976*, Rockwille, 1969.
3. SCHÖNFELD, A., 'Verpackung und ihre Folgen für die Abfallwirtschaft', *Müll und Abfall* **4** (1979) 103.

4. MOSER, W., 'Umwelt und Verpackung unter Berücksichtigung der Wiederverwertung von Verpackungsabfällen'. In: *Symposium Recycling – eine Aufgabe der Zukunft*, Frankfurt, 1975
5. VOGEL, G., *Die Systemsammlung von Altstoffen – Ein Verfahren zur Abfallverwertung*. Ergebnisse des Forschungsprojektes zur getrennten Sammlung von Altstoffen in der Per-Albin-Hansson-Siedlung Ost in Wien 1977–1979, Schriftenreihe des Institutes für Technologie und Warenwirtschaftslehre der Wirtschaftsuniversität Wien, Band 1/1980, Vienna, 1980.
6. DOBIL, M., ESCHRICH, E., *et al.*, *Müllerhebung 1973, herausgegeben vom Österreichischen Bundesinstitut für Gesundheitswesen*, Vienna, 1974.
7. SAMMET, D. and WACHTER, J., 'Verfahrensmöglichkeiten der Gebührenveranlagung bei der Abfallbeseitigung'. In: *Der Städtetag* **29** (1976) 159–163.
8. DOEDENS, H. and GALLENKEMPER, B., 'Einfluß des Behälterraumangebotes und des Behältersystems auf Haus- und Sperrmüllmengen', *Müll und Abfall* **4** (1978) 97–103.
9. Vergl. KERNSTOCK, F.: *Verpackung, Verpackungswirtschaft, Entwicklung und Tendenzen am Beispiel Österreichs und internationale Vergleiche*, Schriftenreihe des Institutes für Technologie und Warenwirtschaftslehre an der Hochschule für Welthandel Wien, Band 1/1975, Vienna, 1975.
10. KOHLHAUSER, G., 'Wandlungen und Prognose der österreichischen Konsumstrucktur'. In: *Monatsberichte des Institutes für Wirtschaftsforschung* **1** (1976) 8–20.
11. *Statistisches Handbuch für die Republik Österreich 1979*, Vienna, 1979.
12. ECOSYSTEM, *Vorstudie Abfallwirtschaftsprogramm der Bundesregierung*, Bonn, 1973, p.8.
13. SPRENGER, W., *Der Einsatz materialsparender und abfallarmer Technologien in der Feinstblechverpackungsindustrie unter besonderer Berücksichtigung der abgestreckten Getränkedose*, Vortragsmanuskript, 1976.
14. Vgl. *Reglementierung Getränkeverpackung*, *Schriftenreihe Verpakkungsforschung*, Heft 10, Vienna, 1979.

4 Returnable and Non-returnable Beverage Containers and Household Waste

Jonathan Fisher

INTRODUCTION AND OUTLINE

The beverage container is currently an issue of topical interest with legislation against the non-returnable being passed in various American states (e.g. Oregon) and Scandinavian countries. The subject is also being reviewed by the EEC and the UK's Waste Management Advisory Council. This chapter presents the results of two related research projects recently undertaken in the UK on this issue.

It first presents some information on the level of environmental impact generated by the alternative returnable and non-returnable beverage containers and, in particular, concentrates on what appears from the literature to be the most important environmental impact of beverage containers: the generation of municipal solid waste. This first project demonstrated the importance of the trippage (the number of times a returnable bottle is filled and used for delivering beverages to the consumer) of a returnable bottle to the current beverage-container debate.

Therefore a second research project was undertaken in order to investigate further this specific area. The data currently available on the trippage rates currently achieved by returnable bottles in the beverage markets of various countries are reviewed, and the results of surveys recently undertaken of consumers' and retailers' attitudes and practices towards returning bottles are then presented. These surveys have revealed much useful and interesting information which gives some indication of the possible factors that could account for the low trippage rates currently achieved in the UK beverage market. Finally, the chapter concludes with an identification of some measures to raise these trippage rates.

THE ENVIRONMENTAL IMPACTS OF BEVERAGE CONTAINERS

Municipal Solid Waste

It is estimated that in the financial year 1977/8 it cost the municipalities about £365m to collect and dispose of the 18 million tonnes of waste generated annually in the UK[1]. The average costs of collecting municipal waste were, in 1977/8, about £14.80 per tonne and the average costs of waste disposal were, in 1977/8, about £5.40 per tonne. The costs of solid waste management have been increasingly rapidly in recent years, as also has the volume of solid waste generated, and these trends are predicted to continue in the future. There is increasing public concern about the environmental impacts of waste disposal and many local authorities face difficulties in finding suitable landfill sites for the disposal of these increasing volumes of household waste.

It had been suggested that beverage containers are the second largest identifiable product category in solid waste. (The largest single product category in municipal solid waste is newspapers, which account for approximately 9% of total municipal waste.) Furthermore, it has been argued that the contribution of beverage containers has been increasing recently in line with the general growth of packaging, which has been identified as one of the major causes of the recent growth in the volume of municipal waste generated.

However, up until now, no definite data were available on the actual size of the beverage container component of the waste stream. Therefore a material balance approach was followed which showed that approximately 520 000 tonnes of waste were generated in 1977 in the UK by containers for carbonated soft drinks, beer and cider. During the research project, 100 analyses of the composition of solid waste were also undertaken in various parts of the UK ranging from Cornwall in the south west of England to Dundee in the north east of Scotland. These analyses, along with the calculations from the materials balance approach, have shown that containers for carbonated soft drinks, beer and cider constitute about 3-4% of the total weight and about 2% of the total volume of domestic waste. In addition, detailed analyses of the composition of solid waste were undertaken at three locations. These detailed analyses revealed that all beverage containers, including containers for spirits, wines, non-carbonated soft drinks (e.g. fruit juices and cordials) and milk and dairy products, as well as the containers for beer, cider and carbonated soft drinks, constituted about 7% of the total weight of municipal waste in the UK.

In the UK milk is mainly distributed door-to-door in returnable bottles, which achieve a very high return or trippage rate. Higginson[2] has estimated that if plastic containers were to be used instead of the returnable glass bottles for the distribution of milk in the UK, an additional 228 000 tonnes of waste would be generated each year. Because of the low density of plastic, the increase in terms of the volume of solid waste generated would be greater.

A comparison of the relative impacts of the major alternative containers for beer, cider and carbonated soft drinks, shown in Table 4.1, indicates that the non-returnable glass bottle generates the most municipal solid waste, followed by the metal can, with the returnable bottle generating the least solid waste.

Litter, Air and Water Pollution, and Energy Consumption

The other environmental impacts of beverage containers (litter, air and water pollution emissions, and energy consumption) have been assessed in earlier published reports and papers[3,4]. The results of the available studies are summarized in Table 4.1. This shows that the returnable system causes less air and water pollution and requires less energy than the non-returnable glass and plastic bottle and the bimetallic and aluminium can systems. These studies showed the metal cans and non-returnable glass and plastic bottles are littered more frequently than the returnable bottle.

Conclusion

It is apparent from the foregoing analysis and from the data presented in Table

4.1 that the returnable bottle has a lower environmental impact than the alternative non-returnable containers for beer and carbonated soft drinks.

TABLE 4.1 The Solid Waste, Air and Water Pollution Emissions Generated by the Distribution of 1000 l of Beer in the Alternative Container Systems

From data presented in *Resource and Environment Profile Analysis of Nine Beverage Container Alternatives*, EPA USA, 1974; Basler and Hoffman, *Studie Unwelt und Volkwirtschaft: Vergleich der Umweltbelastung von Behaltern aus PVC, Glass, Blech und Karton*, 1974, report prepared for the Swiss Federal Office for the Protection of the Environment, Zurich; Research Triangle Institute, *Energy and Economic Impacts of Mandatory Deposits*, 1976, report prepared for the Federal Energy Administration, USA.

Environmental impact	Beverage container							
	Returnable bottle				Non-returnable bottle		Can	
	1 trip	5 trip	10 trip	20 trip	Glass	Plastic	Bimetallic	Aluminium
Solid waste								
Wt (kg)	1056	211	106	53	704	?	148	63
Volume (m^3)	2.85	0.57	0.28	0.14	1.9	?	1.3	0.55
Air pollution(kg)								
EPA study	—	24.0	11.3	8.45*	31.3	28.8	26.6	38.7
Basler and Hoffman study	—	—	—	7.59	145.6	147.2	127.5	?
Water pollution(kg)								
EPA study	—	8.3	4.2	3.3*	6.8	8.2	4.1	7.1
Basler and Hoffman study	—	—	—	0.87	6.05	8.5	14.88	?
Energy (Therms)								
RTI	177	70	51	50	114	115	122	187
EPA	n/a	113	57	42	170	170	142	198

*The EPA data relate to a 19-trip returnable bottle. The three returnable bottles systems analysed in the EPA study required different quantities of supplementary packaging. The difference between the absolute figures for pollution emission given in the two studies is caused by differences in the assumptions and methodologies adopted by the separate studies. (For a more detailed discussion of these studies, see Fisher[13]).

TRIPPAGE

The Importance of Trippage

This favourable environmental impact ranking of the returnable bottle depends essentially on the trippage achieved by the returnable bottle. Therefore this aspect will now be examined in some depth.

Generally, the individual returnable bottle is heavier and more expensive than a non-returnable container, and the returnable system involves greater retailing and distribution costs. However, each time a bottle is returned, savings are achieved in the financial costs associated with the acquisition of the raw materials and their manufacture into a new container, and, hence, in the total costs of beverages in returnable bottles. This is shown diagrammatically in Fig. 4.1. These American data[5] show that a returnable bottle has to be used about five times for filling beverages in order to supply consumers with packaged beer at a lower cost, in purely financial terms, than the non-returnable bottle and can.

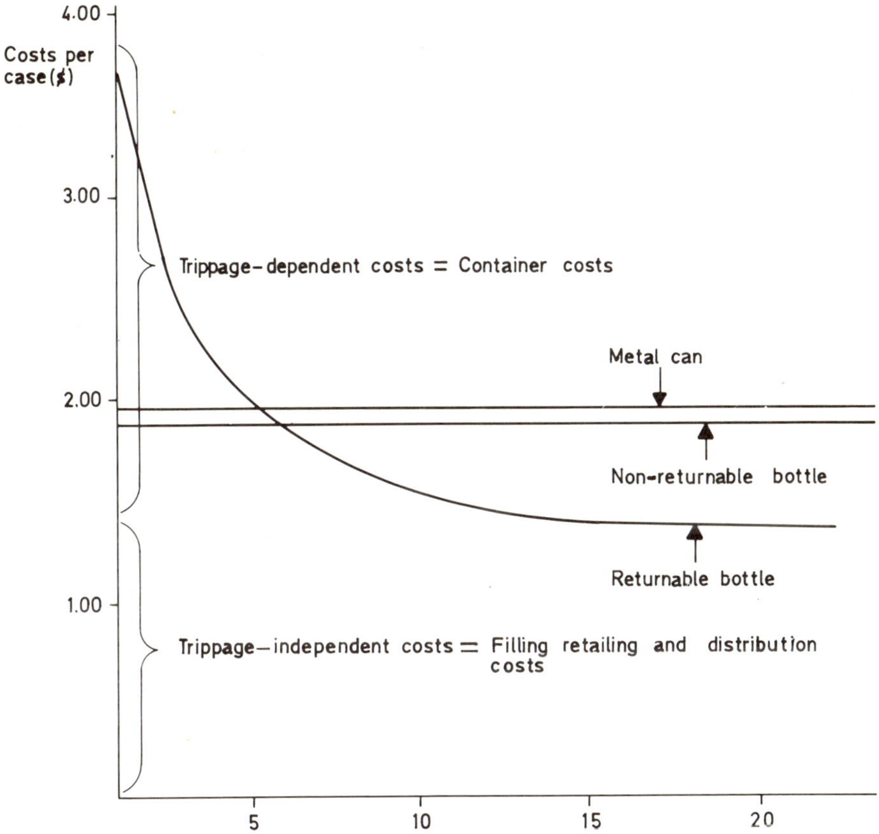

Fig. 4.1 Comparison of the total financial costs of supplying packaged beer in non-returnable containers and returnable bottles with different trippages. Note: trippage-dependent costs are those costs that will vary with the trippage rate achieved, e.g. the costs of the individual container (bottle) which will be amortized by the number of trips achieved by that bottle. The trippage-independent costs are those costs which are invariant of the trippage rate and are incurred regardless of whether a new or a returned bottle is used (e.g. filling, retailing and distribution costs). Stern et al. (Data from[5]).

Similarly, it is evident from Table 4.1 that increases in the returnable's trippage will yield reductions in the environmental impacts resulting from the distribution of the beverages to the consumer. Table 4.2 shows the breakeven trippages that the returnable bottle has to achieve in order to consume less energy and

TABLE 4.2 Break-Even Trippages (T^*) for the Returnable Bottle to have a Lower Environmental Impact than Three Non-returnable Container Systems for Beer and Soft Drinks

Derived and adapted from data presented in Research Triangle Institute, *Energy and Economic Impacts of Mandatory Deposits*, 1976, Federal Energy Administration, USA; USEPA, *Resource and Environmental Profile Analysis of Nine Beverage Container Alternatives*, 1974, EPA, USA; data and information supplied by Metal Box Market Research Division.

Non-returnable container system	Energy	Environmental impact			
		Solid Waste		Air Pollution	Water Pollution
		Weight	Volume		
Non-returnable bottle	2	1.5	1.5	1.2	7.1
Can — bimetallic	1.5-4	7-8	2-3	1.6	16.6
Can — aluminium	1-2	17-20	5-6	1.0	1.0

generate less solid waste, water and air pollution than the non-returnable container systems.

Thus the determination of whether the returnable bottle is the socially preferable container, both in economic and in environmental terms, will depend crucially upon the trippage achieved by the returnable bottle. Therefore there is a need to determine the trippage rates currently achieved in the UK beverage markets and, more importantly, for an analysis of the determinants of trippage and, hence, of how trippage can be raised. The benefits of increasing trippage are the promotion of a viable returnable system that will lower the costs of supplying the consumer with packaged beverages and yield savings in the environmental pollution associated with the acquisition of the raw materials, their manufacture into new containers for beverages and the ultimate disposal of the containers in solid waste.

Review of Current Information on Trippage

UK Data

The trippage rates in the UK beer, cider and soft drinks markets have been declining in recent years and at present are approximately 13 for beer and cider and 9 for carbonated soft drinks.

These figures exceed most of the breakeven trippages (given in Fig. 4.1 and Table 4.2) required for the returnable bottle to be the private and socially preferable container.

However, these are aggregate trippages of the whole beverage market, which contains two sectors: 'on' premise sales (e.g. in pubs or door-to-door sales) and 'off' premise sales (e.g. through off-licence and other retail outlets). These two sectors are distinctly different and, therefore, merit separate consideration, especially since any beverage container policy will primarily be concerned with the 'off' premise sector. The 'on' premise sector represents a closed loop return system where the return of the bottles is easy and straightforward so that the trippage achieved in this sector is much higher (about 30) than that achieved for 'off' premise sales where the beverages are consumed away from the place of purchase and the consumer then has to return the bottle to this place.

However, very little information is currently available on this important area of 'off' premise trippage. A survey was recently undertaken of the proportion of the returnable bottles that consumers currently return to 18 'off' premise retail outlets in Leicester. (A more detailed analysis of the methodology and results of this retailers' survey is given by Fisher and Horton[6].) This survey yielded fairly similar findings to an opinion survey of retailers commissioned by Metal Box Ltd. Both of these surveys, being based on the opinions and estimates of a limited number of retailers, may not yield perfectly accurate data on this subject. Therefore, they can only give a tentative indication that the 'off' premise trippage for carbonated soft drinks and cider may be currently about 5-6 and for beer about 3-5. Nevertheless, these findings are consistent with the estimate for 'off' premise trippage of 4 reported by the UK government's Waste Management Advisory Council[7].

Overseas Experience

The trippage rates currently achieved for returnable bottles in a number of countries are given in Table 4.3. In Norway, Denmark and Sweden, bottles are returned at the rate of 99%, 99.4% and 96% respectively. This yields consumer trippage rates (T_C)[8] of 100, 150 and 25 and, when allowance is made for the breakage rate of the bottles of 1-3%, yields the total trippage rates (T_T)[8] shown in Table 4.3. These data refer to aggregate trippage rates for the whole beverage market ('on' and 'off' premise sales). Unfortunately, data for 'off' premise trippage are, at present, available in only two of these countries: Sweden and Ontario, Canada (for beer). Here, consumer 'off' premise trippage rates are estimated to be 17 and 25, respectively.

This is in marked contrast to the position in the UK where, as has just been shown, the 'off' premise consumer trippage rates are, at present, about 4. Thus Cawdell, in Friends of the Earth's minority report to the Waste Management Advisory Council[9] states (p.7) that '. . . there is not a single country in Europe with beer or carbonated soft drinks trippages less than Britain'.

These apparent distinct differences between the trippage rates currently achieved in the various countries do raise the interesting question of what are the determinants of trippage and, in particular, what are the reasons for the UK's low 'off' premise trippage rate.

Review of Current Information on the Determinants of Trippage

Little research or in-depth analysis had previously been undertaken on the determinants of trippage, especially 'off' premise trippage. The reason conventionally given for the low return rate for bottles is that consumers 'cannot be bothered to return bottles'. Increasing the level of the deposits has usually been the main prescription adopted in the past to raise trippage. The ineffectiveness of this conventional wisdom diagnosis and prescription is evidenced by the fall in trippage rates that has occurred in recent years in the UK in spite of large increases in the deposit levels. This has resulted in increasingly

TABLE 4.3 The Trippage Rates Achieved in Various Countries

Data compiled from: *Beverage Containers: Re-use or Recycling*, 1978, OECD, Paris; *Float Equity Systems for Standard Refillable Soft Drinks Containers*, 1976, Waste Management Advisory Board, Ontario; A. Peaker, 'Resource Saving from the Reintroduction of a Returnable System of Beverage Containers: A Case Study of Experience in Oregon', *Resources Policy* **1**, No. 5, September 1975, pp. 266-276; *Study of the Effectiveness and Impact of Oregon Minimum Deposit Law*, 1974, Applied Decisions Systems Ltd., USA; *The Beverage Market in Sweden, 1973, 1974 and 1975*, PLM Metal Division and Glass Division.

Country	Trippage (T_T)	
	Beer (+ cider)	Soft drinks
UK	13	9
Canada		
Ontario		
Total Market	33	13
Urban Centre		5-7
Quebec	33	
USA	14	10
Oregon	6-20	12-24
Switzerland	60-80	20-70
Germany	25	9
Sweden	17	
Finland	30	
Norway	35	
Denmark	31	

large sums of money being lost by consumers in the form of foregone deposits on their returnable bottles. It is estimated that in 1977 British consumers lost about £18 million in deposits on the 267 million beer, cider and soft drinks bottles that were not returned.

Thus, there was a need for a thorough and wider ranging study of the determinants of trippage, and therefore a research project was set up in Leicester to study this area by obtaining information from the three parties involved in the flow of returned empty bottles from:

(a) consumers
to
(b) retailers
to
(c) bottlers/brewers

The objective of this study was to find out why bottles are at present not returned and to consider what could be done to increase the return rate for bottles. The study consisted of four parts:

(a) a survey of 1485 householders in Leicester and Blackburn and 43 schoolchildren in Colwyn Bay,
(b) a survey of the opinions and practices of 69 retailers in Leicester and Blackburn,
(c) a practical bottle-returning exercise in which a random sample of eight returnable bottles was taken back to 24 retail outlets and the characteristics of the bottles rejected and accepted by each of these retailers noted,
(d) personal interviews with five local and national bottlers.

The Consumer Survey

The Consumer Survey was undertaken in seven separate areas in each of the cities of Leicester and Blackburn and covered the major different types of housing areas (council houses, terraced, semi's and detached houses).

The first section of the survey contained some specific questions concerning each of the beverage bottles shown in Figs. 4.2-6. The consumers were asked:

(a) Do you buy any of these bottles?
(b) Can you tell me whether the bottles shown in these photographs are returnable or non-returnable, or if you do not know?
(c) Of the returnable bottles that you buy, what proportion do you actually return?
(d) Do you know what the deposit level is on each of these returnable bottles shown in these photographs?

The replies to questions (b) and (d) are presented in Tables 4.4 and 4.5.

The surveys found that 6% of the householders who had purchased the returnable bottle in question thought that it was a non-returnable, and a further 7% did not know whether the bottle was returnable or non-returnable. Of particular interest are the replies from those consumers who did not return the returnable bottle. In over a third of the replies from this group, the bottle was not correctly identified as being returnable, with 22% regarding this 'returnable' bottle as being non-returnable so that it is hardly surprising that these consumers had failed to return the bottle.

29% of those householders who had purchased the returnable bottle replied that they did not know what the deposit level was on the bottle; over 80% of the consumers who did not return the bottle said that they did not know what the deposit level was on the bottle, or gave a deposit level that was less than the actual deposit on the bottle.

Thus it is evident that consumers at present have an apparent lack of knowledge about the returnable bottles currently in use in the UK beverage market and their deposit values.

Many of the non-returnables have the words 'No Deposit No Return' embossed on the bottle or prominently printed on the label (see, for example, bottles 9 and 17 in Figs. 4.3 and 4.4). In contrast, many of the returnable bottles in the UK do not have any such identifying feature to inform the consumer that this bottle is a returnable with a refundable deposit on it. Two of the soft drinks bottles and one of the cider bottles selected for this survey did

1 2 3 4

Fig. 4.2 Soft drinks bottles used in the consumer survey in Leicester. Note: in the actual servey, 10 x 8 inch colour enlargements of these photographs were used, so that the consumer was presented with fairly life-size picture of the bottles.

5 6 7 8

9 10 11 12 13 14

Fig. 4.3 Soft drink bottles used in the consumer survey in Blackburn. Note: 10 x 8 inch colour enlargements were used in the survey.

15 16 17 18

Fig. 4.4 Cider bottles used in the consumer survey in Blackburn and Leicester. Note: colour enlargements were used in the survey — 7 x 5 inch in Blackburn and 10 x 8 inch in Leicester.

TABLE 4.4 Householders' Replies on Returnability of Bottles

Data from a consumer survey recently undertaken in Leicester and Blackburn. The sixteen returnable bottles in the survey were bottles 1, 4, 5, 8, 10, 12, 14, 15, 16, 18, 19, 21, 22, 23, 25, and 26 in Figs. 4.2 — 4.6.

Reply	Householders who had purchased the returnable bottle (%)	Householders who did not return the bottle (%)
Returnable	87	64
Non-returnable	6	22
Don't know	7	14

TABLE 4.5 Householders' Replies on Deposit Levels

Data from a consumer survey recently undertaken in Leicester and Blackburn. The sixteen returnable bottles in the survey were bottles 1, 4, 5, 8, 10, 12, 14, 15, 16, 18, 19, 21, 22, 23, 25, and 26 in Figs. 4.2 — 4.6.

Householder's reply concerning the deposit level on the returnable	Householders who had purchased the returnable bottle (%)	Householders who did not return the bottle (%)
Don't know	29	52
Actual level	39	31
Actual level	32	17

have such an identifying feature on their label or the cap. The consumers in the survey more often correctly identified these bottles as being returnable, and more often specified correctly the appropriate deposit levels on these bottles than they did for the returnable bottles that did not have any such identifying feature.

Data were also obtained, from the retailers survey and the bottle-returning exercise, on the information that the consumer receives when he buys the beverage. The retailers surveyed were asked how they currently marked the prices of the beverages they sell in returnable bottles (either on their sales shelves or on the bottles themselves), i.e. if the price of a pint of beer in a returnable bottle was 35p and the deposit on the bottle was 6p, then do they now mark:

(a)	the (gross) price of the beverage *including* the deposit	41p
(b)	the (net) price of the beverage *excluding* the deposit	35p
(c)	the (gross) price of the beverage, inclusive of the deposit, but with the existence (and level) of the deposit on the bottle displayed as well	41p (incl. 6p deposit)
(d)	the price of the beverage and the deposit on the bottle separately	35p + 6p (deposit)
(e)	Any other marking system	

In the bottle-returning exercise, information was obtained on which of these price-marking systems was adopted by the retailer by a visual inspection of the

19 20 21 22

Fig. 4.5 Beer bottles used in the consumer survey in Leicester. 10 x 8 inch colour enlargement were used in the survey.

23 24 25 26

Fig. 4.6 Beer bottles used in the consumer survey in Blackburn. 7 x 5 inch colour enlargements were used in the survey.

retailers' bottles and shelves. The retailers' questionnaire survey and bottle-returning exercise yielded similar results. These two data sources showed that, while 12 of the retailers did identify the deposit separately, most (55) marked only the gross price of the beverage and did not indicate the existence or level of any deposit on the bottle. Presumably, this would affect the consumer's perception of the returnable nature of the bottle and the level of the deposit on this bottle.

All the householders who had purchased a returnable bottle were then asked whether there was any particular reason why they do not return any or some bottles. Their replies to this question revealed that while some householders could not be bothered to return the bottle or said that it was not worth it, nearly twice as many householders said that they did not return a bottle because they found it difficult to return bottles. (A more detailed analysis is available in the literature[6, 10].)

There were three main ways in which consumers find the returning of bottles difficult: some, especially older people, said that the bottles were heavy and awkward to carry; some consumers did not know that the bottle was a returnable bottle and many did not know where they could return the bottle to get the deposit refund on it; many consumers said that they encountered problems in finding a shop that would accept their returnable bottles. This last factor appears to have been the most important cause of the difficulties experienced by consumers. These consumers said that the retailers had refused to accept a returnable bottle which they had taken to the shop. Not surprisingly, more than half of these consumers gave up the attempt to return the bottle and threw it away.

Further information on this aspect was obtained from the householders' replies to the question: 'How easy or difficult is it for you to return the bottles?' (This question was asked *after* the question 'Is there any particular reason why you do not return this/these bottles?', so as not to result in any bias in the reasons given by the householders for their not returning bottles.) More than one third (38%) of those consumers who did not return bottles said that they found it very difficult or difficult to return bottles. Of the householders who said that they did not return bottles because they did not think it was worth it, four said that the returning of bottles was difficult and three said that it was very difficult to return bottles, indicating that it was not worth it for these consumers to return bottles in these difficult circumstances.

A principal objective of the survey was to assess how trippage, or return rates, could be raised. Therefore, the householders were asked: 'What would encourage you to return more bottles?'. In their replies (Table 4.6), 34 people said that they would return more bottles if there were some positive promotion by the bottlers/brewers and retailers of the importance and social benefits of returning bottles. Apart from a recent 'hand it back' campaign run by some soft drink manufacturers, little has been done to demonstrate to the UK consumer that his returnable bottle is wanted back. This contrasts with the situation in Canada where the Ontario brewers and retailers run an extensive 'bring 'em all back' campaign.

167 consumers said that the increases in the deposit levels would encourage them to return more bottles. However, nearly twice as many (285) said that they would return more bottles if it was made easier for them to return the bottles. This suggests that the ease of return is a more important factor than increases in the deposit levels.

The deposit on a returnable bottle is designed to encourage consumers to return the bottle by acting as a financial inducement to compensate them for the

TABLE 4.6 Householders' Replies to the Question: 'What would encourage you to return more bottles?'

Data from a survey of consumers in Leicester and Blackburn; Fisher and Horton[6]; Mercer *et al.*[10].

Reply	No. of householders giving reply
Bottles + retailers said that bottles were wanted back/advertising campaign	34
Easier to return bottles	285
Higher deposit	167
Uniform deposits	2
Government action	14
Nothing, return all now	97
Nothing	7
Other replies*	29

*These included 'shop not close', 'discount on next purchase as well as deposit refund', 'lottery' and 'if bought more returnables'.

inconvenience of returning the bottle. However, it was noted earlier that over half of those consumers who did not return bottles did not know what the level of the deposit was, and a further 31% gave a deposit level that was less than the actual deposit level at present on the bottles. Therefore, for these consumers, the present deposit levels are not acting as a very effective incentive for them to return the bottle. Furthermore, in Oregon (USA) a higher return rate was achieved by a standard 'certified' beer bottle with a 2-cent deposit than by a non-standard beer bottle carrying a 5-cent deposit. Thus, it appears that making it easier for consumers to return bottles (or, in other words, reducing the inconvenience and difficulty of returning bottles) would have more effect on return rates than further increasing the deposit levels.

This importance of the ease of returning bottles is further substantiated by the experience in the beverage markets of various countries. In countries such as Norway, Denmark and Canada (for beer) where it is relatively easy for consumers to return the standard bottles, which are accepted by any shop selling that beverage, a high return and trippage rate are achieved by these bottles. Whilst in those countries such as the UK and Canada (for soft drinks) where it is more difficult to return bottles to the retail outlets, a much lower trippage rate is achieved (see Table 4.3).

The consumers in this survey mentioned five major factors that would make it easier for them to return bottles to the shops.

(a) More or some information, on the bottle's label or at the shop, that the bottle was a returnable bottle with a refundable deposit, and better knowledge of where the consumers could return the bottle to get their deposit back on it.

(b) If it were easier to carry the bottles back to the shop. In the UK, returnable beverage bottles are usually sold across the counter in individual bottles, and it was mentioned earler that some consumers find these individual bottles awkward to carry back to the retail outlet. In Canada, beer is sold in handy-pack paper carriers in which it is convenient to place the bottles when empty and then return the carrier full of empties to the store. In Scandinavian countries (e.g. Denmark and Norway) returnable crates, which also carry a deposit, are used for the sale of beverages to the customer. It is suggested that this is one of the factors contributing to the high return rates currently experienced in these countries.

(c) Closer shops.

(d) Standardization of bottles.

(e) More consumers, however, said that it would be easier if they could take the returnable bottles back to more shops or to any shop (i.e. standardized acceptance of beverage bottles by the retailers). This last aspect, a greater acceptance of returned bottles by the retailers, was the factor that was most frequently mentioned by the consumers, and would appear to be the most important way in which consumers feel that the return of bottles could be made easier for them.

Retailers

The retailers have a central role in the distribution of beverage containers both to and from the consumer and the bottler/brewer. Hence, it was essential to obtain an input of their opinions into this study. Therefore, the retailers' survey included some questions on the retailers' attitudes and policies concerning returnable bottles.

Many retailers said that they disliked the inconvenience and problems caused by the handling, sorting and storing of returned empties. Nevertheless, some retailers said that they positively supported the use of returnables. This was not only for environmental reasons but also for commercial reasons. These retailers said that the use of returnables attracts customers back into their shops as they return with the empty bottles. This is comparable to many retailers' current practice of giving special discounts on certain products, or loss leaders, in order to attract customers into their store, only that with returnables the customers are also given a sum of money (the deposit on his bottle(s)) which they are then likely to spend in the retailer's shop. At current deposit levels, this sum of money could amount to more than £1 for a dozen bottles successfully returned.

The retailers were then asked whether they accepted all returnable soft drinks/beer/cider bottles that people brought to their store or only certain ones. While few retailers accept *any* bottle, most retailers normally placed certain restrictions on the types of returnable bottles that they will accept back from consumers. These restrictions appeared to vary, depending on the type of consumer bringing the bottle to the store. Thus retailers might accept any re-useable bottles of the same size and shape as those used by their bottlers from a regular or bona fide customer (i.e. someone who was likely to buy something), but would not accept these returnable bottles from children. This information from the retailers regarding their acceptance of bottles confirms the earlier findings of the survey of consumers.

Further detailed information on this aspect was obtained from a bottle-returning exercise that was also undertaken. In this exercise, eight randomly selected returnable bottles were taken to 24 shops in the survey areas of Leicester. 130 of the bottles were rejected by the 24 retailers, and 62 bottles accepted. Furthermore, for 10 out of these 62 bottles that were accepted by the retailers, the deposit given was less than the deposit refund that the retailer should have been given for these bottles. Thus, on average, the retailers accepted 2—3 and rejected 5—6 of the returnable bottles in each return exercise. A comparison of the characteristics of each batch of 8 bottles with the characteristics the beverages sold by the retailer revealed some interesting information concerning the rejected bottles. The five categories for these bottles were:

(a) 17 rejected bottles were of a different brand, shape and size to those sold by the retailer. The retailer could not have been expected to accept these bottles.

(b) 34 of the rejected bottles had special distinctive shapes. The retailers sold the same beverage in the same size of bottle but in bottles of a different shape and brand name. Standardization of bottle shapes might help to increase the trippage and life of these bottles.

(c) 6 of the rejected bottles were also a different shape from those sold by the retailers. However, in these cases, the rejected bottles were standard, but the retailers sold the same beverage in the same size of bottle, but in bottles of a different distinctive shape. Standardization of all bottle shapes might also help to increase the trippage of these bottles.

(d) 46 of the rejected bottles, however, were of the same size and shape as the beverage bottles sold by the retailer but were of a different brand name or label (i.e. standard one-pint beer bottle not being accepted by an off-licence which sold the standard pint beer bottles of another brewery). Standardization of bottle shape would have no effect on the trippage rate achieved by these bottles, under this current situation concerning retailer acceptance of returned bottles.

(e) 9 rejected bottles were the same size, shape and brands as those sold by the retailers. The retailer said that he would not accept three of these bottles because, even though he stocked the same brand (in fact Guiness), 'the bottle was not from my bottler' (the bottler's name was in small print on the bottom of the label). For the same reasons stated in (d) above, standardization of bottle shapes would have no effect upon the trippage of these bottles.

It is evident that, while a few retailers positively support the use of returnables since this attracts back into the shop customers who are likely to buy something, most of the retailers place some restrictions on the types of returnable bottles that they will accept from the customers. The two main reasons given by these retailers for stipulating the restrictions were that their supplier would not accept the bottles and because the returned bottles caused problems (e.g. storage) for them. These retailers were unwilling to perform the service of accepting returned empty bottles for which they did not perceive any compensation or rewards. Thus, there appears to be a marked reluctance to accept back empty bottles on the part of many retailers who stipulate these restrictions so as to reduce the number of returned bottles (and the concomitant problems) that they have to handle.

Brewers and Soft Drinks Bottlers

The brewers and soft drinks bottlers want as many as possible of their returnable bottles back, since this re-use saves them from having to buy a new bottle. Therefore, some bottlers and brewers have taken measures in an attempt to increase the return rate ot their bottles. The National Association of Soft Drinks Manufacturers has recommended the use of a returnability symbol for returnable bottles which has been adopted by some bottlers. Many others do now print words such as 'Please return bottle' on their labels. In addition some bottlers also print on their bottle tops the level of the deposit charged on the bottle (as is evident for bottle 10 in Fig. 4.3). These latter methods of identifying the returnable nature of the bottle and its deposit are more noticeable and probably have a greater impact on consumer awareness than the returnability symbols, which tend to be in rather small print on the bottom corner of the label, as can (just) be seen for bottle 10 in Fig. 4.3. It was pointed out by some brewers that their ability to change their labels was constrained by the costs of redesigning the labels, and that the size of the labels on their (relatively small) bottles limited their ability to include any readily noticeable feature informing the consumer that the bottle was a deposit-bearing returnable. (Nevertheless, it was also noted that bottlers are willing to allocate valuable sales promotion space on their labels to clearly state 'No Deposit, No Return' on the non-returnable bottles sold through supermarkets (as can be seen on bottle 9 in Fig. 4.3).) Separate neck labels would be one effective method of displaying such an identifying feature, although this would involve some increase in the costs of labelling. Another alternative is the embossing of such words as 'Returnable Bottle' or 'Deposit Bottle — Please Return' on the actual bottle (as is currently accomplished on many returnable milk bottles in the UK, and also on the new returnable Coke bottles recently introduced by Coca Cola).

CONCLUSIONS AND POLICY IMPLICATIONS

The first part of this chapter has shown that the returnable bottle generates less municipal waste and has a lower environmental impact than the non-returnable containers. However, this favourable ranking for the returnable depends essentially upon the number of times this bottle is returned and used for filling beverages — its trippage. This has highlighted the importance of the returnable bottle's trippage to the current controversy over returnable vs. non-returnable beverage containers.

From the currently available information and data, it appears that the trippage rates in the UK beer and soft drinks market have been declining in recent years and that they are lower than the trippage rates currently achieved in overseas beverage markets.

The chapter then reports the results of some recent surveys of beverage consumers, retailers and bottlers; these results have some interesting implications for two important questions in the current debate: why consumers do not return bottles and how the return rate, or trippage, of returnable bottles could be increased.

Why Consumers Do Not Return Returnable Bottles

The consumer survey showed that some consumers did not return bottles because they could not be bothered or were unwilling to return bottles. It would have been better, in both economic and environmental terms, if these consumers had purchased a non-returnable in the first place rather than their present failing to return a returnable and, perhaps, if they were then encouraged to recycle these non-returnables (e.g. through Bottle Banks). Nevertheless, it is doubtful whether the consumer for whom it was 'not worth it' to return the bottle to a shop for the deposit refund of between 4 and 10p on each bottle would be willing for no financial reward to take his non-returnables to a bottle bank.

However, the survey also revealed that more consumers did not return the bottle because they found it difficult to do so. These difficulties arose from two major sources: firstly, an apparent lack of knowledge amongst consumers that the bottle was a returnable with a refundable deposit on it; secondly, the problems that consumers encountered in locating a retailer who would accept their returnable bottles. Many consumers said that a retailer had refused to accept a returnable bottle that they had taken to his shop. This finding was supported by the results of the bottle-return exercises in which, on average, five to six out of each batch of eight returnable bottles were rejected by the retailer. The retailers' replies in the questionnaire survey also showed that most retailers stipulated that they would accept only certain types of bottle.

Thus, it appears that the reasons for the British consumer's failure to return bottles are more complex than the simple, conventional diagnosis that consumers 'just cannot be bothered to return bottles'. These surveys also indicate that, rather than just accept these low and declining rates of returning bottles by UK consumers as being inevitable and irremediable, there are some positive measures which could beneficially help to halt these current trends and could even raise trippage rates.

How the Return Rate or Trippage of Returnable Bottles Could be Increased

Raising the deposit levels is the traditional remedy adopted in the past in an attempt to halt declining trippage. However, increasing deposit levels above their current levels would not be very effective at present, since so many consumers either do not know what the deposit is on the bottle or think that it is less than the current deposit level. The deposit is supposed to act as an inducement to compensate the consumer for the inconvenience of returning bottles. Therefore, it would be more effective and efficient to alter this other aspect (the inconvenience of returning bottles) by making it *easier* for the consumers to return bottles.

The Consumers' ease of returning bottles could be improved by:

(a) *Giving the consumers some more information* that a bottle is returnable and has a deposit refund on it. Some bottlers do now put such information on their bottle tops and labels. Other bottlers and brewers should give serious consideration, especially when they are redesigning their labels, to putting some readily identifiable feature on their labels or tops so that the consumer can see that a bottle is a returnable and has a deposit on it. In addition, more retailers should clearly and separately identify the fact that there is a refundable deposit (currently hidden) in the 'gross' price of beverages in returnable bottles.

(b) Standardization of bottle shapes, *on its own*, would not have much effect on increasing trippages of returnable bottles in the present circumstances in the UK beverage market, where it is difficult for consumers to return both standardized and distinctively shaped bottles to the retail outlets. In the bottle-returning exercises, 46 out of the total 130 rejected bottles were of the same size and standard shape as the beverage bottles sold by the retailers.

(c) *Increased acceptance of returned bottles by the retailers* is also required. It has been advocated by Cawdell[9] that this should be achieved by government regulations such as the mandatory deposit refund clauses of the Oregon Bottle Bill. Under such regulations, the retailers and distributors must refund the deposit on any carbonated beverage container if they sell that type of bottled beverage. However, if such a regulation were deemed to be unacceptable or not practicable, the government could instead pass a directive exhorting retailers to accept more returned bottles from consumers. Alternatively, the retailers could be encouraged to accept more bottles by the adoption of a differential deposit system by which the retailer would receive a handling premium for the returned empty bottles accepted back from the consumer and sent back to the bottler who is thereby saved from having to buy more new bottles.

The results of the surveys reported in this paper have shown that, if retailers were to accept more returned empty bottles, this would ease the major difficulty that is currently experienced by British consumers in returning (or rather attempting to return) their bottles. Some, or one, of the above possible measures for achieving an increased acceptance of returned bottles by retailers would be a more effective and efficient method of increasing the trippage and life of a returnable bottle than would further increases in the deposit levels.

REFERENCES

1. CHARTERED INSTITUTE FOR PUBLIC FINANCE AND ACCOUNTANCY (CIPFA) (1978), *Financial, General and Rating Statistics: 1977/78* Statistical Information Service, CIPFA, London.
2. HIGGINSON, A.E. (1971), *The Composition, Yield and Analysis of Domestic Refuse*. Paper presented at the 73rd Annual Conference of the Institute of Public Cleansing, June 1971.
3. OECD (1978), *Beverage Containers: Re-use or Recycling*, OECD, Paris.
4. FISHER, J.C.D. (1978), 'A social Appraisal of Beverage Containers Recycling: A UK Perspective', *Resources Policy*, Vol. 4, No. 2, June 1978.
5. STERN, C. *et al.* (1975), *Impacts of Beverage Container Legislation on Connecticut, and a Review of the Experience of Oregon, Vermont and Washington State*. Department of Agricultural Economics, University of Storrs, USA.
6. FISHER, J.C.D. and HORTON, P. (1979), 'The Return of Returnable Bottles: A Study of Consumers' and Retailers' Attitudes and Practices concerning Returnable Bottles', *Resources Policy*, Vol. 5, No. 4, December 1979, pp. 279 — 297.
7. WASTE MANAGEMENT ADVISORY COUNCIL (1980), *Study of Returnable and Non-returnable Containers*. Report of the Packaging and Containers Working Group of the Waste Management Advisory Council (WMAC), London.
8. The total trippage rate (T_T) is derived from the following formula:

$$T_T = \frac{1}{1-(ARR-APB)}$$

Where ARR is the proportion of bottles returned (by customers), and APB is the proportion of bottles broken at the bottling plant, retail outlet, or in transit. (This breakage rate is expressed as the proportion of bottles broken at these stages in the course of supplying a certain level of beverage sales in returnables.)

The consumer trippage rate (T_C) is derived as follows:

$$T_C = \frac{1}{1-(ARR)}$$

9. CAWDELL, A. (1980), *Study of Returnable and Non-Returnable Containers*, Minority Statement of A. Cawdell (Friends of the Earth) to the Packaging and Containers Working Group of the Waste Management Advisory Council.
10. MERCER, D., NG, M., LOWTHER, C. and CHAPMAN, J. (1980), *Blackburn Waste Survey: A Study of Public Attitudes to Household Waste Collection and Disposal*, Profit from Waste Ltd., Manchester Business School, July 1980.
11. APPLIED DECISION SYSTEMS LTD. (ADS) (1974), *Study of the Effectiveness and Impact of the Oregon Minimum Deposit Law*, Salem, Oregon, USA.
12. BASLER and HOFFMAN (1974), *Studie umwelt und Volkwirtshaft: vergleich der Umweltbelastung von behaltern aus PVC, glass, Blech und Karton*. Report prepared for the Swiss Federal Office for the Protection of the Environment, Zurich.
13. FISHER, J.C.D. (1979), 'A Study of the External Costs of Beverage Containers', Public Sector Economics Research Centre, University of Leicester, Leicester.
14. LIDGREN, K. (1976), *Containers and Environmental Policy*, Student-Literatur, Lund, Sweden.
15. PEAKER, A. (1975), Resource Savings from the Re-introduction of a Returnable System of Beverage Containers; A Case Study of Experience in Oregon. *Resources Policy*, Vol.1, No. 5, September 1975, pp. 266-276.
16. RESEARCH TRIANGLE INSTITUTE (1976), *Energy and Economic Impacts of Mandatory Deposits*, Report prepared for the Federal Energy Administration, USA.
17. US EPA (1974), *Resource and Environmental Profile Analysis of Nine Beverage Container Alternatives*, EPA, USA.
18. WASTE MANAGEMENT ADVISORY BOARD (1976), *Float Equity Systems for Refillable Soft Drinks Containers*, Waste Management Advisory Board, Ontario Ministry of the Environment, Toronto, Canada.

Jonathan Fisher obtained a degree in Economics, then spent three years undertaking research in Environmental Economics at the Public Sector Economics Research Centre (University of Leicester). During this research, he wrote a number of articles on various environmental questions including returnable vs. non-returnable beverage containers, the recycling of waste materials and pollution taxes. Currently, he is an administrator at the Environment Directorate of the Organisation for Economic Cooperation and Development, where his work includes economic analyses of such evvironmental issues such as the costs of pollution control and the assessment of environmental damages.

This paper is based on research performed at the Public Sector Economics Research Unit, University of Leicester, UK.

The opinions expressed in this chapter are the author's own and do not necessarily reflect the views of the OECD.

5 The Economics of Different Refuse Transport Systems

Jaakko Paatero

INTRODUCTION

More careful planning of waste disposal systems and adoption of more purposeful and practical public cleansing systems make it possible to use integrated refuse disposal systems. By refuse disposal systems is meant, in this context, the entire method by which refuse is transferred from the place where it originates to the place where it is finally returned to the industrial or ecological cycle. Fig. 5.1[1] illustrates a number of different waste disposal systems. The refuse is taken from its place of origin to the local collection and transport point near the house, where it is temporarily stored, then by way of the actual transport stage to the place where it is treated, which may, for example, be a refuse dump, from where it is finally returned to the natural ecological cycle. Refuse to be recycled into industrial processes is first sorted. Fig. 5.2 shows the different ways in which the collection and transport of refuse may be organised.

Analysis of waste disposal as a functional whole, from the place of origin to its final destination, results in the kinds of activities shown in Fig. 5.3. It is clear that in some cases certain stages of the waste disposal process may be omitted, while other stages may be combined in different ways. In most cases, however, a waste disposal system can be studied in two different sections: the first consists of the collection, storage and transport stages, and the second of the treatment of the refuse, its use and final disposal. The following factors are of importance in such an approach to the problem of waste disposal.

(a) The collection and transport of the refuse comprises a considerable part of the whole waste disposal process since between 60 and 80% of the total costs involved in waste disposal are spent on the collection and transport stages. Rationalisation of the choice of collection and transport methods therefore have a marked effect on the efficiency of the waste disposal system and the total economic cost.

(b) The collection and transport of refuse constitutes an important link in the disposal chain and must function perfectly in all conditions. The containers or bins used for the intermediate collection and storage of refuse must be adapted to the vehicles used for transporting the refuse, and the system coordinated into a well functioning whole.

(c) The treatment of refuse, its use and final disposal are often kept quite separate from the collection and transport stages of the disposal process.

(d) The development of refuse and collection systems and the smooth operation of these play an important part in the visible disposal of waste. The treatment of refuse and the ways in which it is finally disposed of, on the other hand, are not always seen as part of the overall town picture. Nonetheless, these are reflected in the town surroundings and in the total economic cost of waste disposal. The further development of treatment and disposal methods often come to the fore only when the collection and transport stages operate smoothly.

Although a waste disposal system can, in principle, be studied in two separate sections, it should nevertheless be borne in mind that the way in which the refuse is treated may have a very significant effect on the choice of collection and transport methods.

REFUSE COLLECTION AND TRANSPORT SYSTEMS

Refuse collection and transport systems are integrated systems of equipment and methods by which refuse is transferred from the places where it originates to the sites where it is treated or pretreated. The ways in which refuse is collected and transported in a house or block of flats often form a distinct and separate stage from the stage where it is transported to the treatment site. In such cases the collection of refuse is usually limited to the collecting of household waste in the kitchen, placing it in bins or bags and carrying it from the kitchen to dustbins or refuse containers in the yard. In most places, newspapers and magazines are collected separately and taken to a different container intended specifically for this kind of refuse.

In large blocks of flats, etc., there may be special chutes to facilitate the transport of refuse. However, in recent years the building of such chutes has not been very popular, partly for economic reasons and partly because experiences with such chutes have not been entirely good. The reason has largely been poor planning: not enough attention has been given to the transport of refuse away from the collection points at the lower ends of the chutes. In Sweden, however, there is increasing use of such refuse chutes[2]. Present Swedish legislation lays down that residences with more than two homes and more than two floors must be equipped with a refuse chute if the refuse cannot be handled in some other satisfactory way (SBN 1975, 43:11). A chute need not be built, however, if it leads to a deterioration of working conditions for those employed in the removal of refuse, if it involves considerable extra cost or if there are other good reasons, such as building restrictions or cultural or environmental reasons. In general, it should be noted that Sweden's new refuse disposal legislation places the working conditions of refuse collectors above the comfort of residents. It is possible that such legislation may also be introduced in Finland in the near future.

As far as blocks of flats and residences are concerned, attention has been largely directed towards the collection of refuse from homes and its transport away, rather than internal methods of collection. However, it is worth noting that systems for the internal collection and transport of refuse are economically important, if technical solutions to waste disposal are to be found.

The collection of refuse and its transport away from homes is the most visible stage of waste disposal. Moreover, numerous systems have been developed for this stage of the entire process. All the systems involve different kinds of bins, containers and vehicles to transport them. When comparing the costs of collecting and transporting refuse, it is most usual to take into account only the cost of collecting refuse from houses and transporting it away to the municipal refuse site, etc. The cost of internal collection and transport is disregarded even though it may differ considerably according to the system in use within the

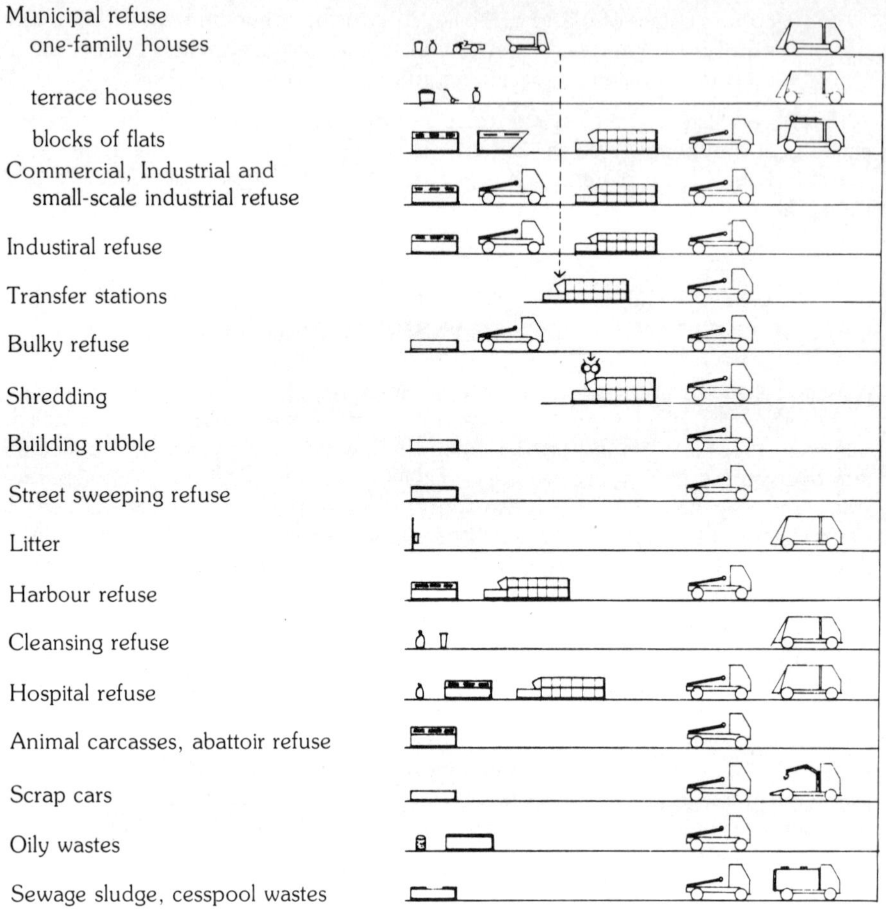

Fig. 5.1 Various activities in municipal waste management.

house. An obvious example is the use of different kinds of chute or pipeline arrangements.

Refuse collection and transport methods can be divided into four main classes: bin methods, demountable container methods, transfer loading methods and pipeline methods. Bin methods may involve the emptying of the bins or the exchange of full bins for empty ones, and the containers used may be of a permanent or disposable type, e.g. sacks. Transfer loading methods are two-stage methods of transport, and may be used with either bin methods or demountable container methods. Pipeline methods are also two-stage transport methods; the first stage is the actual pipeline system and the second stage involves the use of a demountable container. It should be noted that there are innumerable variations in each main class.

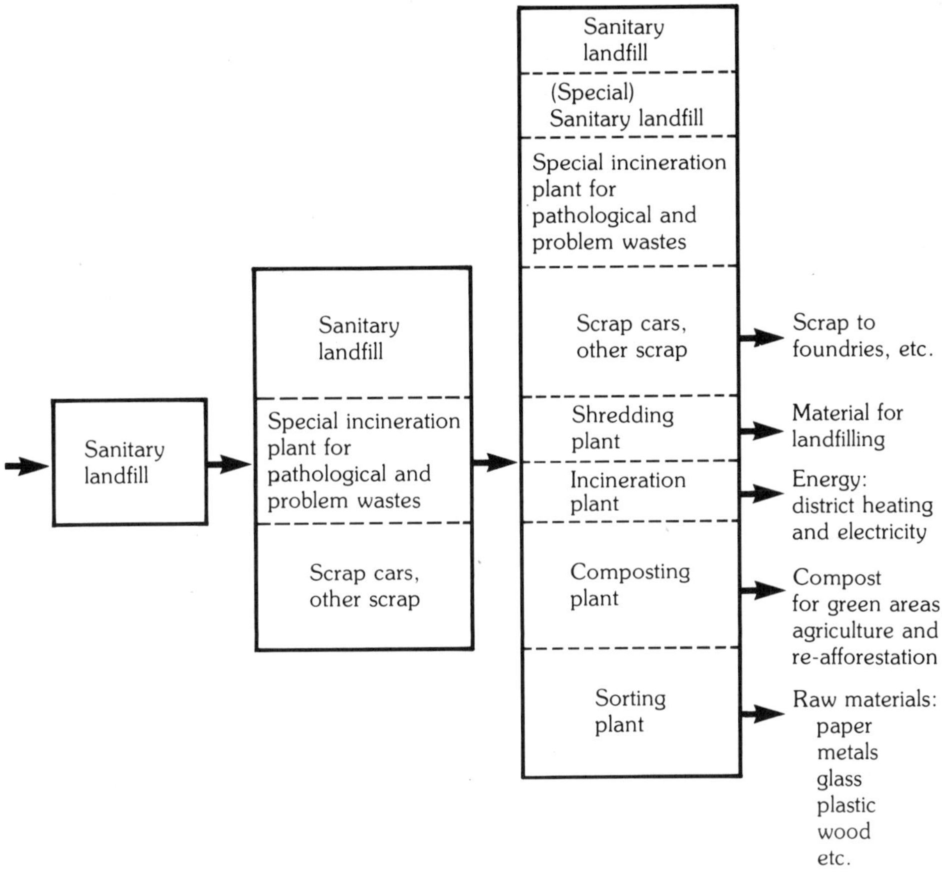

Bin Methods

Refuse Bin/Truck Equipped with Boom Crane

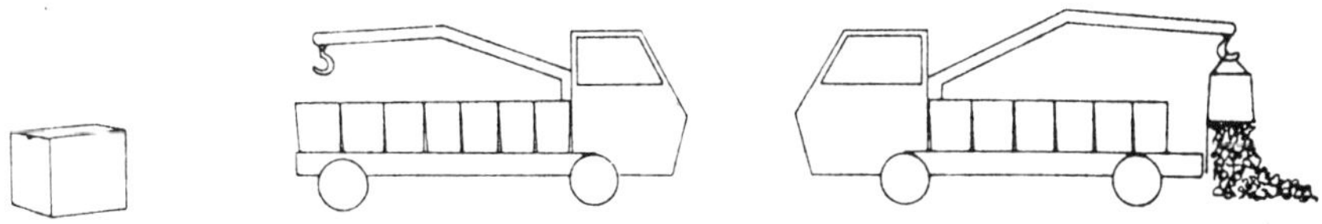

This method may be of the exchange type, in which the full bin is exchanged for an empty bin by means of the truck crane, or it may be the emptying method, in which the bin is hoisted onto the platform of the truck to be emptied and then returned to its original place. The refuse is transported either in containers on the truck platform or in the truck's load space, and emptied from the truck either by means of the crane or by tipping the truck platform. This method is no longer used to any great extent as it has been replaced by more economical methods. Nonetheless, it is still in use in the Soviet Union, for example. In Finland it is used mainly for the collection of waste paper.

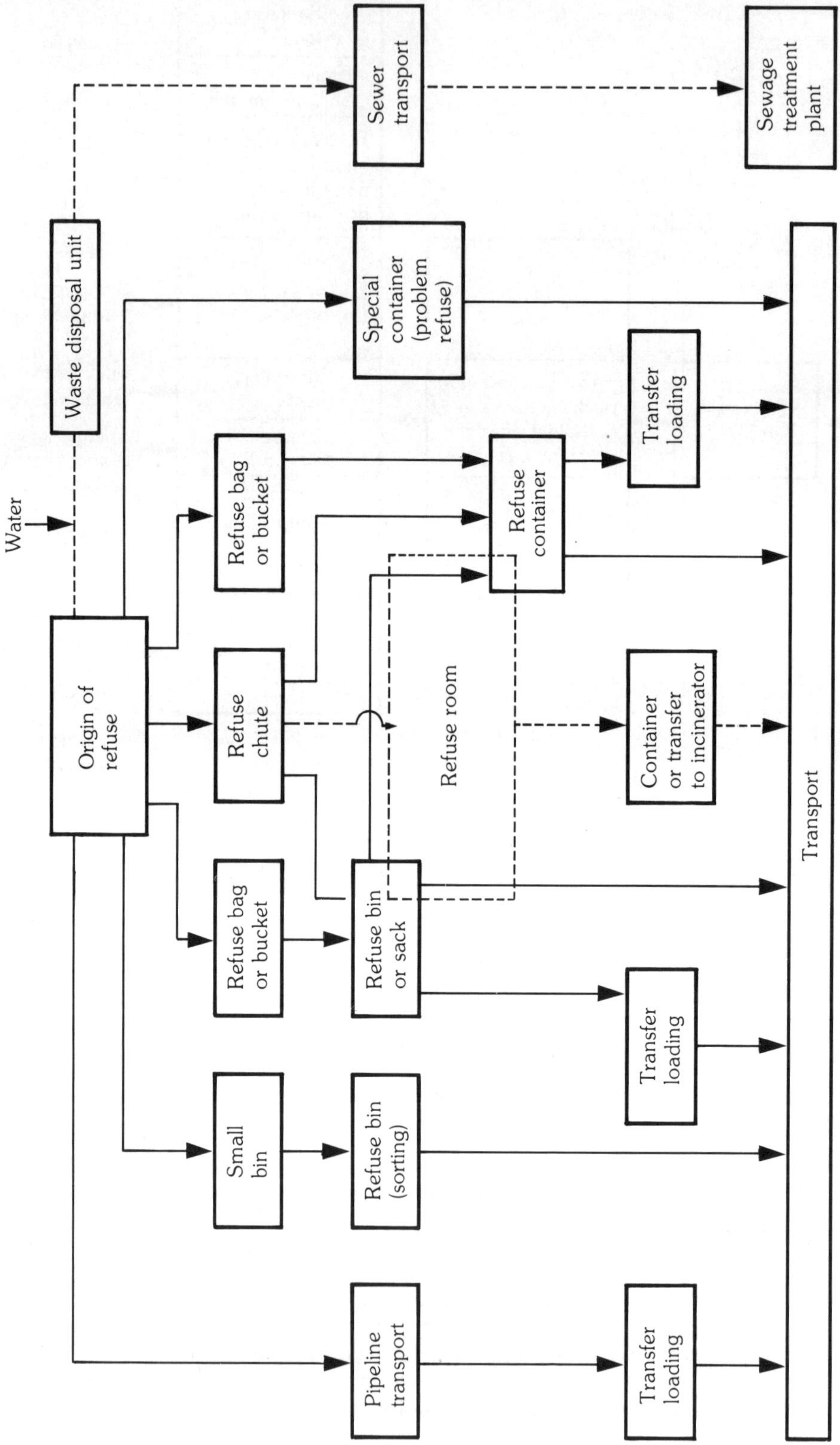

Fig. 5.2 Systems of waste disposal in residential blocks.

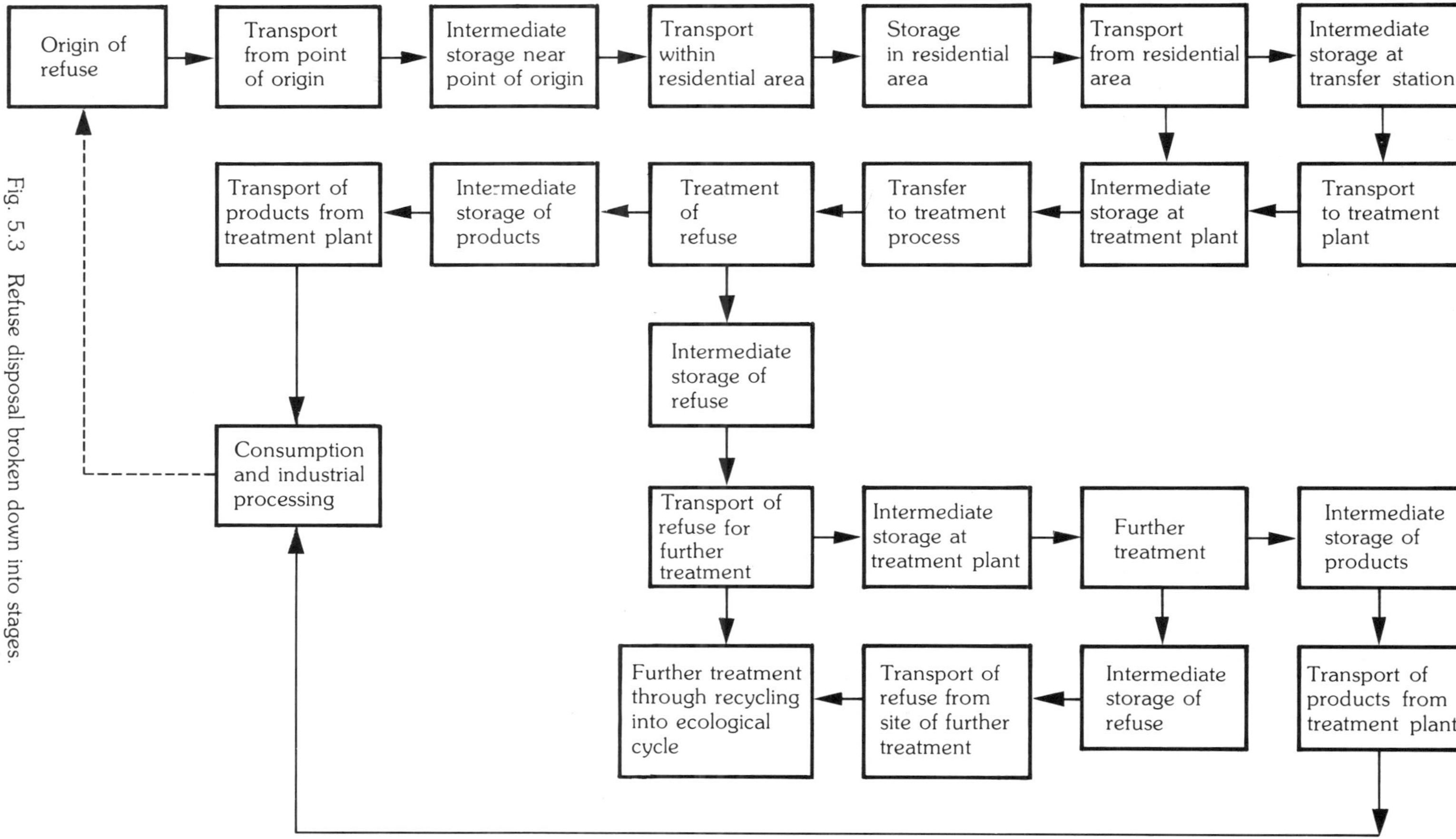

Fig. 5.3 Refuse disposal broken down into stages.

Refuse Bin/Compacting Refuse Collection Vehicle

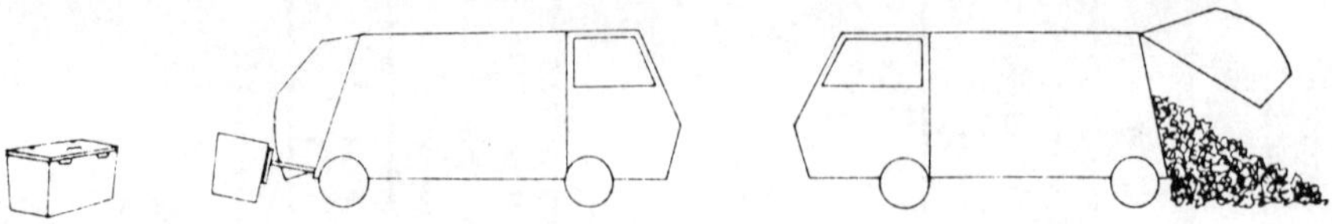

The full bin, sack or small container (1-5 m^3) is emptied either manually or mechanically into the loading hopper of the vehicle. The refuse is then compressed by means of a compacting plate or screw device into the loading space of the vehicle. Recent developments in Central Europe have led to the use of small containers (1-5 m^3) as an intermediate step between bins (with a volume of less than 1 m^3) and large containers (with a volume of more than 5 m^3). Small containers have not so far been widely used in Finland but in West Germany, for example, good results have been achieved by combining the emptying of 5 m^3 containers with the collection of cumbersome refuse[3].

Vehicles equipped with compacting apparatus can transport about 50-80 m^3 of loose refuse at a time, depending on the size of the vehicle's load-carrying space and the efficiency of the compacting equipment.

This method is in general use throughout the developed world for the collection of household refuse and of industrial refuse similar to household waste. In some cases vehicles of this type are also used for collecting sorted waste paper.

Refuse Sack/Light Collection Vehicle

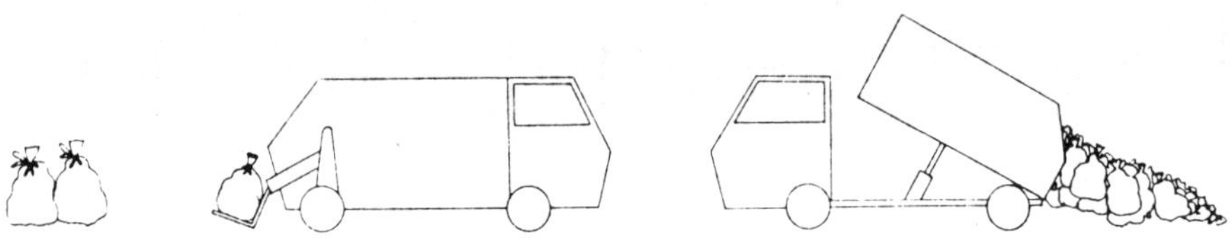

The full refuse sacks are collected by means of a mechanical loading device and a light vehicle equipped with a loading platform that can be tipped. Small vehicles loaded manually belong to the same class, as also do bins used with mechanically or manually emptied light collection vehicles. Vehicles of this type can transport about 5-10 m^3 of loose refuse at a time. There are even vehicles on the market with a load capacity considerably less than 5 m^3.

The method is used principally for the collection of household waste. The refuse load is subsequently often transferred to a compacting refuse vehicle, to a transport container equipped with compacting apparatus or to a permanent transfer station.

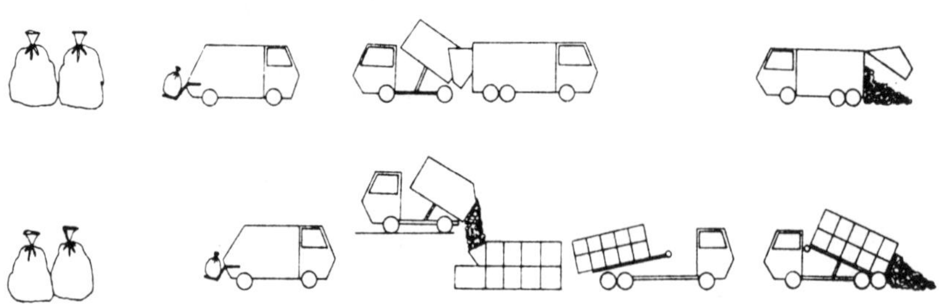

The refuse can also be transported to a nearby refuse dump or treatment site, however, in which the method is then a typical bin method (using disposable containers). Small collection vehicles along these lines are being developed throughout the developed world for use on the satellite principle, and special attention is being paid to the mechanisation of collection.

Container/Compacting Refuse Collection Vehicle with Winch

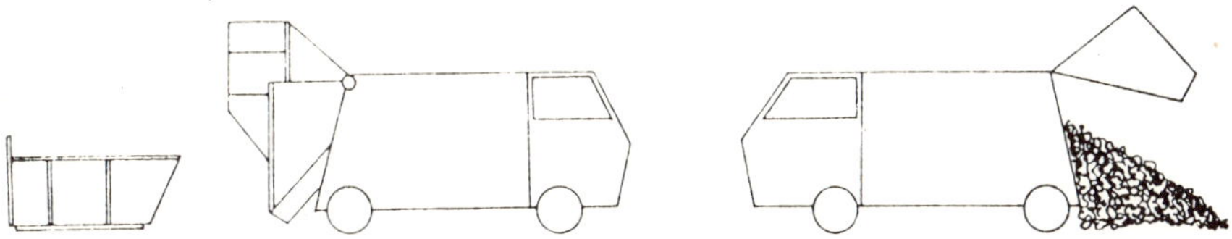

The full container (usually 6-8 m^3) is emptied mechanically by means of a winch arrangement into the loading hopper of a compacting refuse vehicle and the refuse is compressed into the load-carrying space of the vehicle. A compacting collection vehicle of this type can transport 50-80 m^3 of loose refuse, depending on the volume of the load-carrying space, the efficiency of the compacting equipment and the weight of the vehicle's own mechanical apparatus.

The method has become very widespread of late, since it can be used for emptying bins, sacks and containers. Such a type of strong multipurpose vehicle can also be used for transporting various types of cumbersome household refuse, such as furniture, household appliances and machines. Recent labour safety legislation has been carefully observed in the design of the latest models of refuse collection vehicle.

Demountable Container Methods

Refuse Container/Truck with Demountable Body Equipment

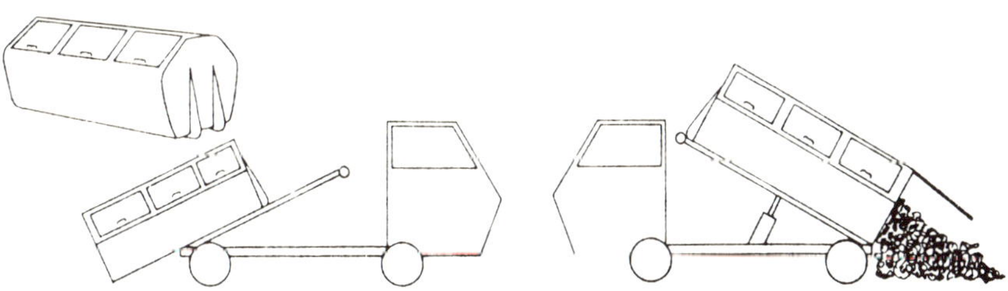

The full refuse container is transported by means of a truck equipped with demountable body apparatus to the refuse treatment site or transfer station. The container is emptied by tipping. A container not equipped with compacting apparatus can transport 5-20 m^3 of loose refuse, depending on the volume of the container and the type of refuse. Since the weight of household refuse is only 80-100 kg per m^3, the effective load is relatively small despite the large size of the containers. In addition to the Multilift demountable container system, the Normacont and Traillift systems in use in Finland are also systems of this type; the latter two are used principally with tractors. In addition to forklift truck systems, mention should also be made of lift dumper, skip loader and load lugger systems in use in Sweden and Central Europe; these are not used in Finland.

This method is used for the collection of both household and industrial refuse. The lift dumper, etc., systems are used principally for collecting and transporting building refuse. The Multilift system is also used for this type of refuse; when this is the case, open steel platforms are used for initial collection of the rubbish.

When closed refuse containers are used for collecting household refuse, they are used mainly to collect refuse from houses where 1-20 m^3 accumulates daily and where the journey to the treatment site does not exceed 15 km.

Compacting Container/Truck with Demountable Body Equipment

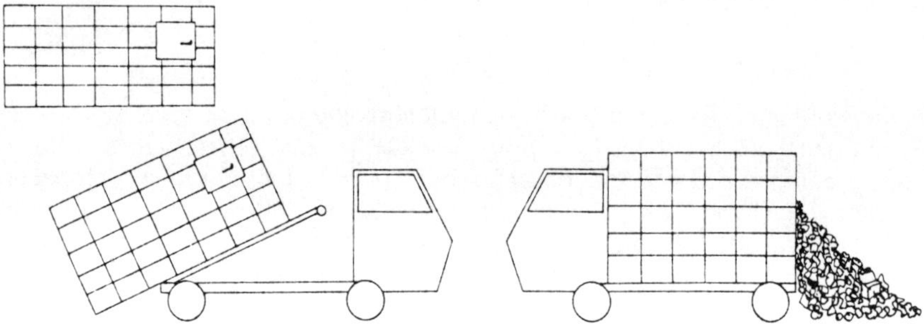

The refuse is collected directly or emptied from refuse containers into a container equipped with plate or screw compacting apparatus. The full container is then transported by means of a truck with demountable body equipment to the refuse treatment site. The container is emptied either by tipping or by means of a hydraulically operated unloading device. The compacting apparatus may be fixed to the outside or the inside of the container. Such a container can transport 40-80 m^3 of loose refuse, depending on the volume of the container and the efficiency of the compacting apparatus.

The method is widely used for collecting commercial and industrial refuse and also for collecting refuse from houses where the accumulation of refuse is 10-50 m^3 each day and where the journey to the refuse treatment site does not exceed 30 km. Variations of this method have been developed for handling refuse from ships. Compacting containers can also be equipped with chutes at the lower end. However, for technical reasons, it is recommended that only fixed refuse compactors be equipped with chutes.

Refuse Compactor/Refuse Container/Truck with Demountable Body Equipment.

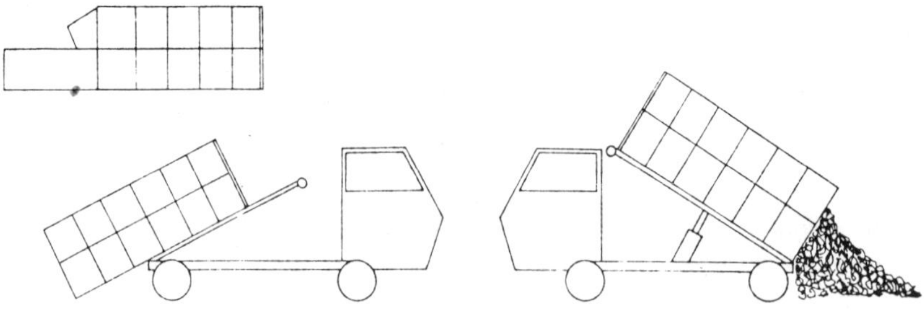

The refuse is collected directly or emptied from large containers into the loading hopper of a refuse compactor of the packer-plate or feed-screw type. The loose refuse is then compressed into a detachable refuse container attached to the compactor. The full container is then detached and transported by means of a truck with demountable body equipment to the refuse treatment site, where the container is emptied, usually by tipping. Such a container can transport 50-100 m^3 of loose refuse, depending on the volume of the container and the efficiency of the compactor. The compression rate is usually 1:4 - 1:6 in the case of household or commercial refuse.

Because of the versatility of their use, compactor/container systems are available in a wide range of types, ranging from small compactor units to large transfer-load types. A further reason for the wide range of available types is the manufacturers' firm conviction in the marketability of a wide range; a conviction which has not always been justified.

Recent recognition of the possibilities of using different vehicles as a source of power for a separate compactor has also led to new developments. The container is equipped with only a packer plate or piston and the necessary controls, while the truck is equipped with an extra hydraulic cylinder and the requisite locking devices. The extra hydraulic cylinder takes power from the vehicle's hydraulic system and transfers the necessary power to the refuse container by means of the controls. One advantage of this system is that no source of electricity is needed.

The compactor/container system is used mainly for collecting commercial and industrial refuse and for collecting refuse from households producing 10-100m^3 of refuse a day and where the distance to the refuse treatment site is no more than 30-40 km.

Transfer Loading Methods

Transfer loading methods involve two stages of transport, where the refuse is transferred from one form of transport to another. Because of the great variety of refuse-handling equipment at present available there are many different transfer loading systems[4]. A survey of transfer load stations in Finland has recently been completed and it must be admitted that they are few in number and small in size[5].

Usually only transport by road is considered when talking of transfer loading. However, there are other methods of transfer loading in use involving the transfer of refuse from road transport to rail or water transport[6]. In addition to rather large transfer stations of the conventional type[7], in recent years new light transfer stations have been developed; these may be either mobile or permanent stations. The principle of transfer loading methods is illustrated in Fig. 5.4.

Because of the desire to reduce the number of refuse dumps, a system has been arrived at comprising a number of transfer stations and one very large, efficient refuse tip. In such a system the transfer stations function to some extent as mechanical refuse dumps which, provided they are properly operated, do not pollute the environment or give rise to the side effects associated with refuse dumps. These transfer stations can be operated at relatively low cost. Furthermore, they offer considerable advantages: they provide the residents of an area with a refuse-reception centre; the local authority public cleansing department derives economic benefits; and refuse arrives at the final refuse treatment site in large quantities. The principles of such transfer stations are illustrated in Fig. 5.5[8].

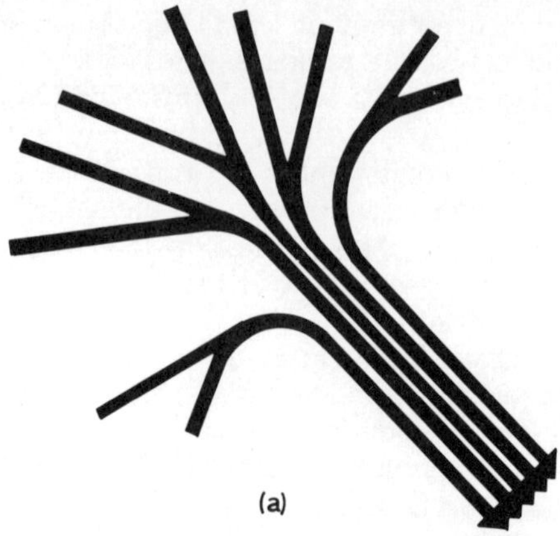

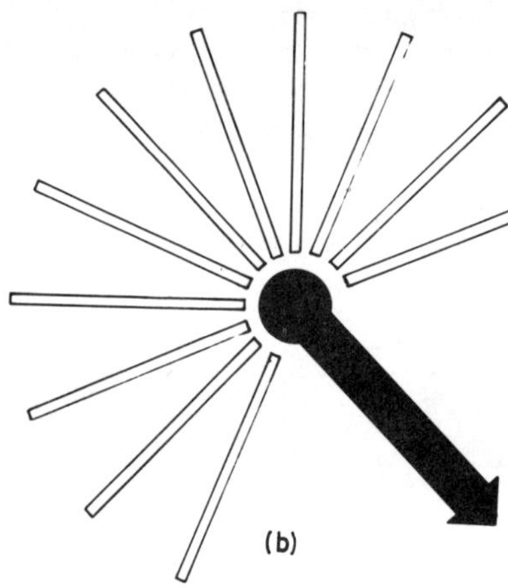

Fig. 5.4 Diagram illustrating principle of transfer loading. (a) The refuse is transported straight to the treatment plant. (b) Refuse is taken first to the transfer station and then to the treatment plant.

Concrete evidence of the practical advantages of such transfer stations compared with a rubbish tip is that a refuse compactor can be successfully used at an easily run station to crush refuse in a concrete hod and feed the crushed and compressed refuse straight into an open transport container, making use of height differences. The full tansport containers can then be driven away to the final treatment site.

Refuse Container/Collection Vehicle/Transport Container/Truck with Demountable Body Equipment

The full refuse container is transported by means of a collection vehicle equipped with special high-lift demountable body apparatus, for example, to an

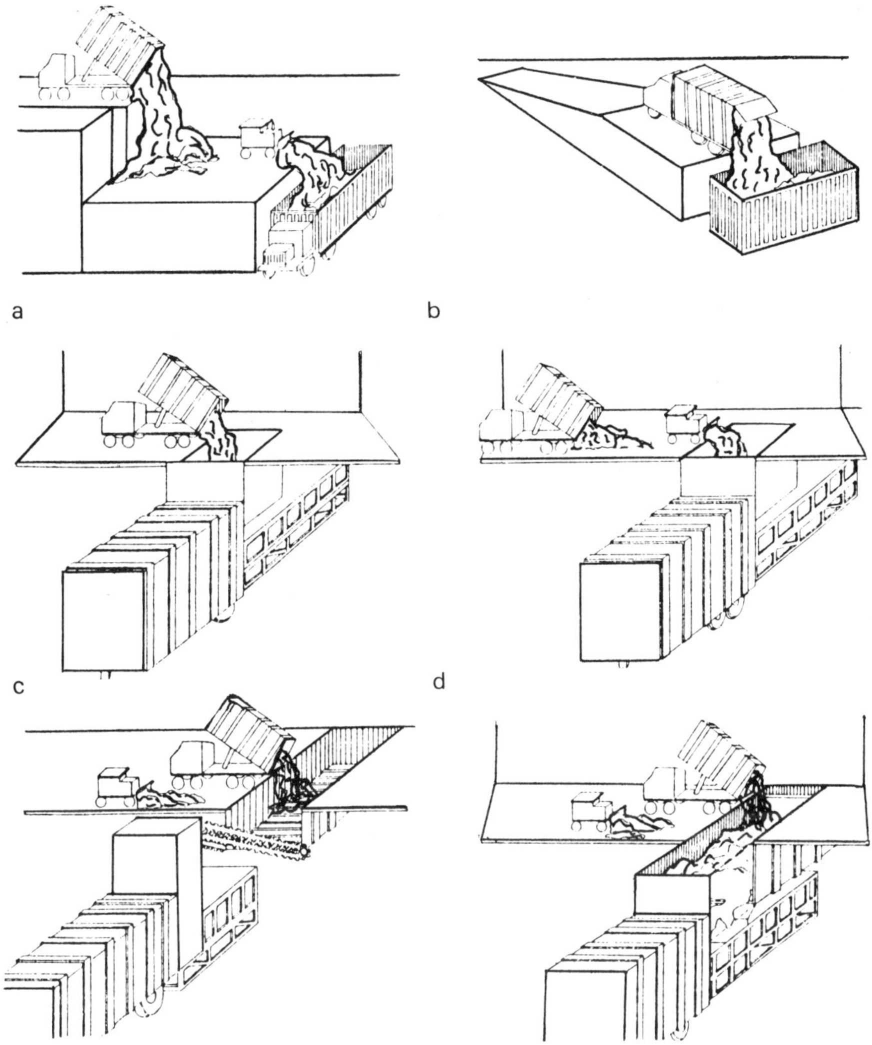

a b c d e f

Fig. 5.5 Different types of mechanised refuse tip solutions[8]. (a) Refuse emptied straight on to storage platform. (b) Refuse emptied into storage container. (c) Compactor application: refuse emptied into loading hopper of compactor. (d) Compactor application: refuse emptied on to intermediate storage platform. (e) Compactor application: refuse emptied on to adjustable feeding conveyor. (f) Compactor application: refuse emptied into loading hopper equipped with packer plate. *(Reproduced from Solid Wastes Management* **22**(12) (1978) 28, by permission).

open transport container into which the refuse is emptied. This is usually done by tipping, making use of the difference in height made possible by the high-lift apparatus. An important factor here is that the transport container is not equipped with a compactor; instead the refuse is transported onwards uncompressed. When the transport container has been filled with refuse, the load is covered with a protective tarpaulin or a roof on the container. The refuse container is then transported by a truck equipped with either a special high-lift device or an ordinary demountable body to the refuse treatment site. With an open transport container it is possible to transport 30-40 m^3 of loose refuse at a time, depending on the size of the container and the type of refuse. A variation of this method is to use a natural or specially constructed difference in height so that the refuse can be tipped from the collection vehicle into the transport container. In such a case high-lift apparatus can be dispensed with. The transport container may consist of a semi-trailer, which is then towed to the treatment site by any suitable tractor. The semi-trailers are emptied by means of a hydraulically operated unloading plate or by opening a band transporter in the floor of the transport container. In some systems the transport containers can be emptied by tipping the whole vehicle using special tipping devices.

The method can be made more effective by crushing the refuse before it is loaded into the transport container, by means of either an ordinary refuse shredder or a refuse crusher operating in a separate concrete hod. As the crusher travels forwards and backwards it breaks up the refuse in the hod and forces it straight into the open transport container by means of a packer plate. The transport container is positioned beneath the lower end of the hod, using the difference in height.

Refuse Container/Truck Equipped with Special Apparatus/Covered Transport Container

Full refuse containers are transported by means of a vehicle equipped with a high-lift demountable body device to the loading hopper of a specially built covered transport container into which they are emptied. The refuse is emptied into the transport container by tipping the load platform. The transport container has compacting equipment and controls but no power unit. The power required for compressing the refuse is taken from a hydraulic cylinder attached to the transport vehicle; this is connected to the vehicle's own hydraulics. The truck is backed up to the transport container and the hydraulic cylinder is attached to the compressing apparatus of the transport container by means of special locking devices. When the transport container is full (it can transport 50-60 m^3 of loose refuse), it is driven away by a truck with high-lift or ordinary demountable body equipment to the refuse treatment site to be emptied, usually by tipping. This method is very useful in residential areas where covered refuse containers are already in use for collecting refuse. An advantage of this method is that the transport container can be moved to practically any transfer loading place. The choice of the transfer loading will depend, however, on environmental factors.

Refuse Bin or Container/Collection Vehicle/Compactor/Transport Container/ Transport Vehicle

The full bins or containers are emptied into the collection vehicle, which transports the refuse to a permanent transfer station where it is emptied into the loading hopper of a compactor. Such a hopper may be a cone, hod, silo, etc., with the necessary controls. The refuse containers can be transported to the transfer station by any suitable vehicle. At the transfer station the refuse is compressed by means of a plate or screw compactor into the transport container. When it is full, the transport container is detached from the compactor and driven away to the treatment site. This can be done by means of a truck equipped with demountable body apparatus, to which in some methods a trailer may also be attached, or by a tractor to which the transport container in the form of a semi-trailer is attached. Permanent transfer stations are normally equipped with container transfer and exchange machinery and this is often highly automated. Fig. 5.6 shows the system used at a large transfer station in the Gothenburg local authority area.

The transport containers used in this method are capable of transporting 100-150 m^3 of loose refuse at a time; the amount depends on the size and shape of the container, the type of refuse and the efficiency of the compactor. Conventional transfer stations are used for the rational transport of large quantities of refuse and also in many instances as a substitute for small local refuse dumps to some extent. A fairly general trend at present is to reduce the number of refuse tips and at the same time to improve the operation of remaining ones.

A transfer station can also be built in the centre of a large town; for example underground, as has been done in central Stockholm[9], where about 50 loads of refuse are compacted by emptying refuse vehicles into one of two loading hoppers. The refuse is then compressed by means of special refuse compactors into 30 m^3 transport containers. This is probably the only underground transfer station in the world. It has been blasted out of rock and is situated about 15 m below ground level in the very centre of Stockholm. There is storage space for 14 transport containers. When a transport container is full, it is transferred by means of a crane onto a specially built vehicle to which is attached a trailer. This combination, 18 m long, can transport two 30 m^3 transport containers to the refuse treatment site about 25 km away and makes ten two-way journeys each day.

In recent years there has increasing interest in Finland in the transport of refuse from water to land and vice versa. Special transfer handling equipment has been developed by means of which transport containers carried by road on the backs of trucks can be transferred to a specially constructed vessel and then back to trucks for further transport by road. The system is illustrated in Fig. 5.7. This system permits satisfactory handling of refuse in difficult situations by means of road-water-road transport in sea and lake areas.

Pipeline Methods

Pneumatic and screw methods of transport are widely used in industry for carrying different kinds of powdery and fine-grained materials. Since the early 1960's, pneumatic transport has also been used for municipal refuse. In addition to various collection and transport systems for household refuse, pneumatic methods have also been used for district refuse handling in residential areas where the refuse is collected and transported by means of a pipeline to a central refuse collection centre. At the centre the refuse is separated from the transporting stream of air and then usually compressed into transport containers

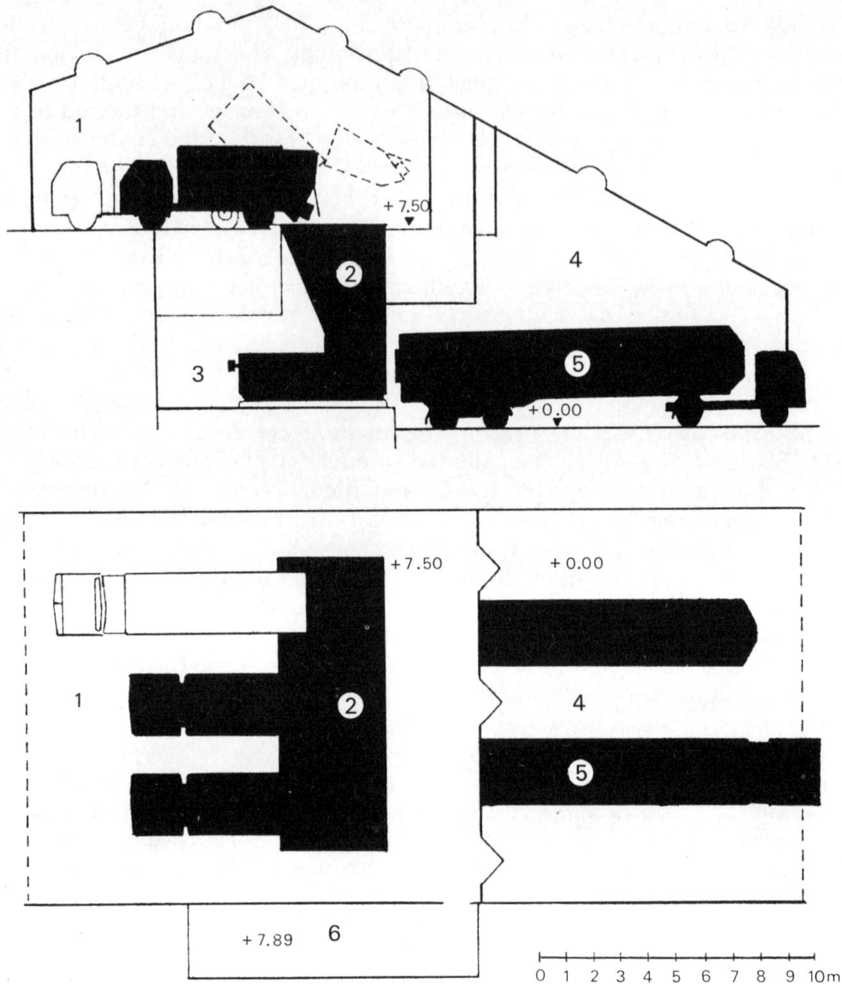

Fig. 5.6 Diagram illustrating operation of transfer station in Gothenburg, Sweden. The refuse is first weighed and the truck is backed up to the emptying station (1). The refuse is emptied into a cone-shaped chute with two separate compartments (2). At the bottom of the chute there are two compactors in the compacting hall (3), which are connected to the two compartments of the chute. In the emptying hall (4) semitrailers (5) are backed up to the compactors, which compress the refuse into the trailers. The whole operation can be supervised from the control room (6), enabling one man to control the whole process.

which are taken away by trucks equipped with demountable body apparatus to the refuse treatment site. The principles of pneumatic pipeline methods are illustrated in Figs. 5.8 and 5.9. The most important users of these methods are hospitals, government offices, industrial plants and heavily built-up residential areas. Nowadays the pneumatic transport of refuse has been developed into a relatively advanced technology offering a genuine alternative in new residential blocks and densely populated areas[10-17]. One considerable advantage that has been increasingly stressed is that the refuse does not have to be touched by human hand at any stage of the collection process.

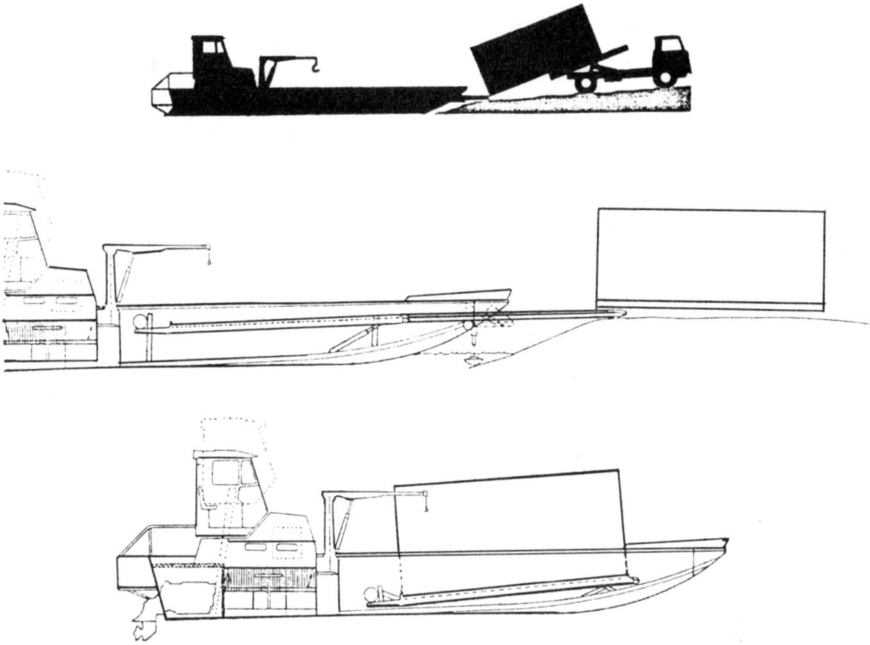

Fig. 5.7 Transport of refuse by means of road/water/road system.

In recent years a new pipeline method has arisen alongside the pneumatic system; this is the screw method (Fig. 5.10)[17]. It has also been used for the collection and transport of refuse in residential areas. The refuse is transported by means of a shaftless screw or spiral revolving in a pipeline which terminates at the refuse collection centre. There the refuse is usually compressed for further transport to the refuse treatment site. In such systems at present in use in Sweden the screw has a diameter of 350 mm and the internal diameter of the pipeline is 460 mm. The screw is operated by means of a motor at one end of the pipe. The power requirement is 1-3 kW for each conveyor[17]. There are at present three screw conveyor systems in operation in Sweden. The maximum length of the screw is considered to be 40-60 m, after which a new screw conveyor needs to be installed.

In addition to the dry pipeline methods described here, there are also systems in which water is used as the transport medium. The oldest of these is the Garchey method[16], in which all household waste is removed by means of a pipe leading from the kitchen for further treatment. This method requires a separating device similar to a septic tank from which sedimented refuse can be removed by suction. One variation of this method is the waste disposal unit installed in the kitchen in which food remains are ground and washed away into the sewage system. Water transport methods have not in general become very widespread because of the high cost factor. Furthermore, in Finland the installation of waste disposal units in kitchens has been met with reservations because of the risk of blockages.

Among the advantages of pipeline methods of transporting refuse are the absence of smells and noise, aesthetic and hygienic benefits and the reduction in

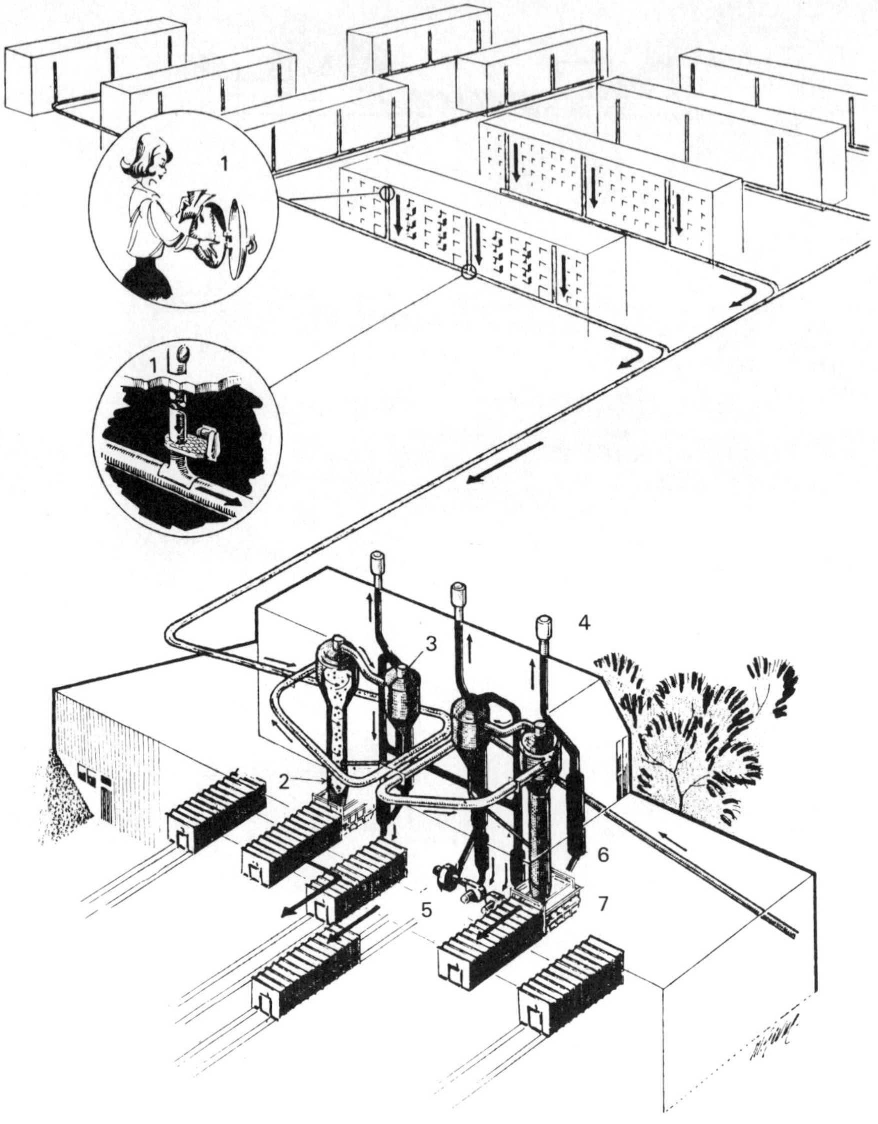

Fig. 5.8 Pneumatic pipeline for transport of refuse: (1) chute feed, (2) large refuse separator, (3) wet refuse separator, (4) Air outlet, (5) high-pressure blowers, (6) silencer, (7) compactor.

traffic. Weighed against the ecological benefits are economic considerations and also technical difficulties, such as pipeline wear and, above all, the danger of blockages. Furthermore, this method is not suitable for all types of household refuse. When considering pipeline methods of waste disposal in Finland, Finnish legislation has to be taken into consideration; this stipulates that in cases where non-conventional methods of waste disposal are used, prior approval of the local refuse disposal authority must be obtained.

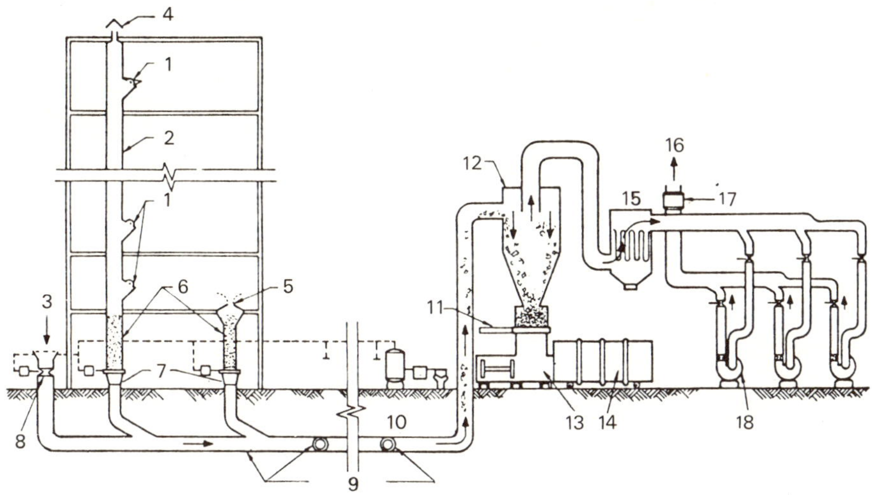

Fig. 5.9 Pneumatic pipeline system for use in residential blocks[10]: (1) chute feed, (2) refuse chute, (3) air inlet, (4) relief inlet, (5) floor-mounted chute feed, (6) refuse storage section, (7) discharge valve, (8) air inlet control valve, (9) pipeline joints, (10) compressor, (11) slide gate, (12) cyclone separator, (13) compactor, (14) transport container, (15) air filter, (16) air outlet, (17) silencer, (18) exhaust blowers. (Reproduced from P.H. Luiten, in: *National Waste Processing Conference 1976*, by permission of The American Society of Mechanical Engineers).

ECONOMIC COMPARISON OF DIFFERENT SYSTEMS

Conventional Systems

The costs of different refuse transport systems consist of the direct vehicle costs resulting from the collection and transport of refuse and also of the costs for bins, containers, etc., needed for the collection of the refuse. The vehicle costs comprise both the capital cost of the vehicles and the operating costs. The effective time that a vehicle can be used is an important factor when comparing the costs of different systems. This time includes the time required for collecting and emptying the refuse bins or containers as well as the time spent in transporting the refuse to the treatment site, emptying it and then driving the empty vehicle back to the collection area. It should also be noted that, in addition to the actual time spent in collection and transport, there are various dead periods involved in refuse collection. In the case where a compacting refuse collection is used for collecting and transporting refuse in bins and sacks, this dead time amounts to almost 13% of the total collection and transport time[18]. Bin and container costs are made up of capital and operating costs. The cost function of a transport system has the following form[1]:

$$K = k_a \cdot \frac{T}{60 \cdot V} + k_k \tag{5.1}$$

where K is the total cost per m^3 of loose refuse, k_a is the cost per hour of vehicle time, T is the time spent in collecting and transporting load (minutes), V is the amount of loose refuse per load (m^3), and k_k is the collection bin costs per m^3 loose refuse.

The cost per hour of vehicle time depends on the capital and operating costs

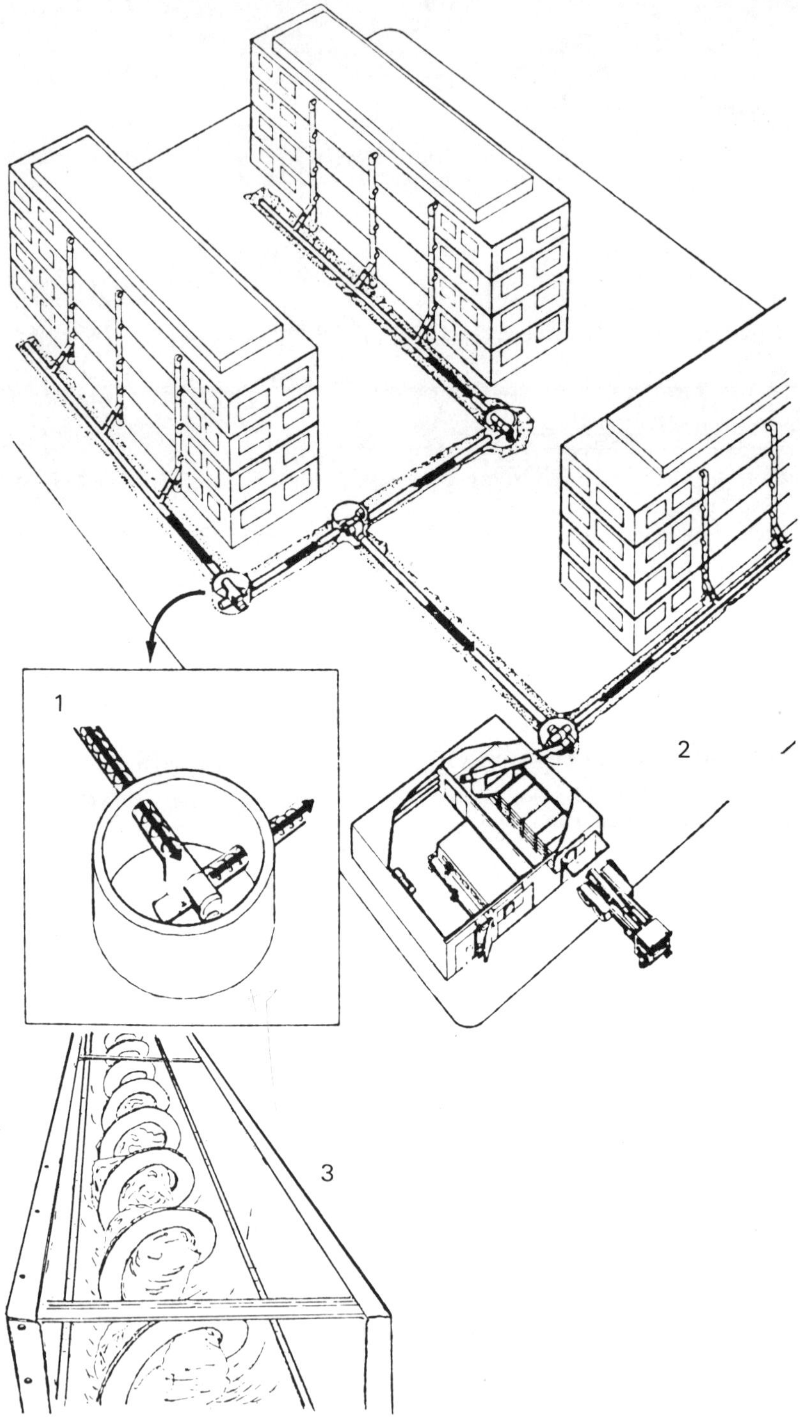

Fig. 5.10 Screw conveyor system[17]: (1) transfer unit, (2) reception centre. (3) spiral installed in pipeline. (Reproduced from L. Bodelins, in: *Proceedings of the 1980 ISWA Congress*, London, 1980, by permission of the International Solid Waste Association).

as well as on the degree to which the vehicle is effectively used and the number of men manning the vehicle. When calculating the capital cost of the vehicle and its equipment and writing it off over short depreciation times, simplified models which do not fully take account of the influence of interest rates can be used[19]. However, it should be noted that such an accounting model is likely to lead to considerable error if depreciation periods of 4-5 years or longer are used. The choice of the correct accounting model is particularly important when calculating the total costs of a capital-intensive refuse transport system. The model used here is a general one in which both wage costs and actual vehicle costs have been taken into account when calculating the hourly costs of the vehicle. Elsewhere these wage and actual vehicle costs have been kept separate[20,21]

The collection and transport time for a full load, T, may be split into two parts:

$$T = t_k + t_s \tag{5.2}$$

where t_k is the collection time based on empirical observation or on time and method studies for a full load — including the time required for emptying the load (min); and t_s is the time for transport of load from collection area to treatment site and return journey (min).

Using an average, empirically observed speed (which is a function of the transport distance) and distance for the transport of the refuse from the collection area to the treatment site, Equation 5.2 may be rewritten in the following form:

$$T = t_k + \frac{120\,s}{v} \tag{5.3}$$

where s is the transport distance from the collection area to the treatment site (km), and v is the average speed (km/h), which in Finland has been observed[22] to be:

s (km)	v (km/h)
<10	35
10-30	40
>30	50

It is clear that Equation 5.3 cannot, in practice, vary in degree with regard to the last term. However, the accuracy obtained should be sufficient for a practical comparison. Fig. 5.11 shows a comparison between empirically observed Finnish and German materials. In the German material, covering an area of 2-60 km, the equation used is[21]:

$$\left(\frac{\bar{v}}{\text{km/h}}\right) = 16 \cdot \left(\frac{s}{\text{km}}\right)^{0.3} \tag{5.4}$$

Equation 5.1 can be written as follows:

$$K = k_a \cdot \frac{t_k}{60\,.\,V} + k_a \cdot \frac{2s}{\bar{v}\,.\,V} + k_k \tag{5.5}$$

or, in more general terms,

$$K = k_1 + k_2 + k_3 \tag{5.6}$$

where k_1 is the indirect transport costs arising principally out of collection of the refuse, k_2 is the actual refuse transport costs, and k_3 is the collection bin cost.

The transport time per load of different systems depends on the density of

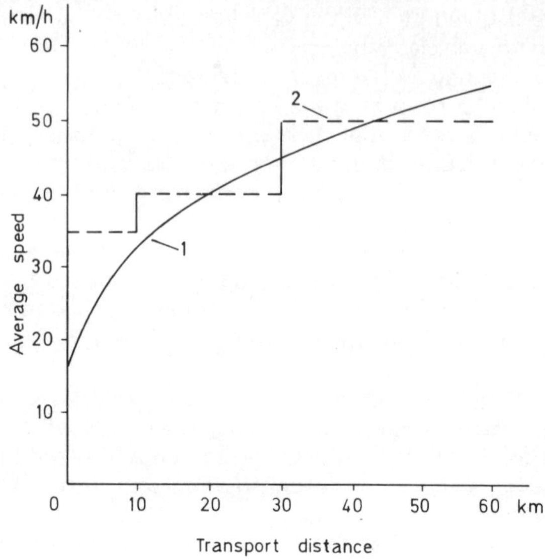

Fig. 5.11 Dependence of average speed $\bar{v}$ on transport distance s. Curve 1 illustrates a function based on German material, $\bar{v} = 16.\ s^{0.3}$, and Curve 2 a simplified model based on Finnish experience.

population, traffic conditions, refuse-handling facilities of the areas from where the refuse is collected, etc. Consequently, it is rather difficult to make generalizations about collection times even though the different stages have been analysed in detail. Time-and-method studies have nevertheless been used to try to chart the economics of different collection and transport systems and for pricing the collection and transport of refuse [18,23,24]. It should be noted that all calculations used for comparing different systems include a number of assumptions and simplifications, which, however, are usually based as far as possible on practice.

Bin costs depend not only on the capital cost of the bins but also on the degree to which they are used and on the usage costs. The volume of the refuse bins should be related to the amount of refuse accumulated and to the collection methods used. If the amount of refuse that accumulates between collection intervals is larger than the volume of the refuse collection unit used, the transfer to two collection units will normally mean an increase in the costs for each cubic metre of refuse. This increase in costs can be evened out by using different sizes of refuse container or bin at the transfer stage. Nevertheless, it should be noted that different types of refuse bin or even different sizes of the same type of bin differ in the degree to which refuse settles in them. As a result, the volume-weight ratio of loose refuse is not constant but varies from one type of bin to another [21]. Therefore, when calculating bin costs per cubic metre of loose refuse, it is a mistake to use constant volume values; in general, this error has been disregarded. When calculating bin costs account should be taken of the maximum degree to which each bin or container can in practice be filled.

In any comparison of the costs of different systems it is important that only the volumes to which the bins or containers are actually filled should be used. In calculating vehicle time costs, however, they may be based on the load-carrying spaces of the vehicles even though they are not fully utilised for the collection of refuse.

Fig. 5.12 shows the costs per cubic metre of loose refuse for ten different transport systems calculated as a function of the amount of refuse accumulated at a single collection point. In the calculations it has been assumed that the maximum quantity of refuse per bin or container is about 80% of the nominal volume of the refuse container used. It has also been assumed that the refuse collection vehicles are in use 8 h/day and that the distance between collection point and treatment site is 10 km in Fig. 5.12a and 30 km in Fig. 5.12b. In the calculations shown in Fig. 5.12, ideal utilization of bins has been assumed, i.e. that only completely full bins are emptied. If the bins were not full (80% full), this would mean extra expense per cubic metre of loose refuse in terms of both bin

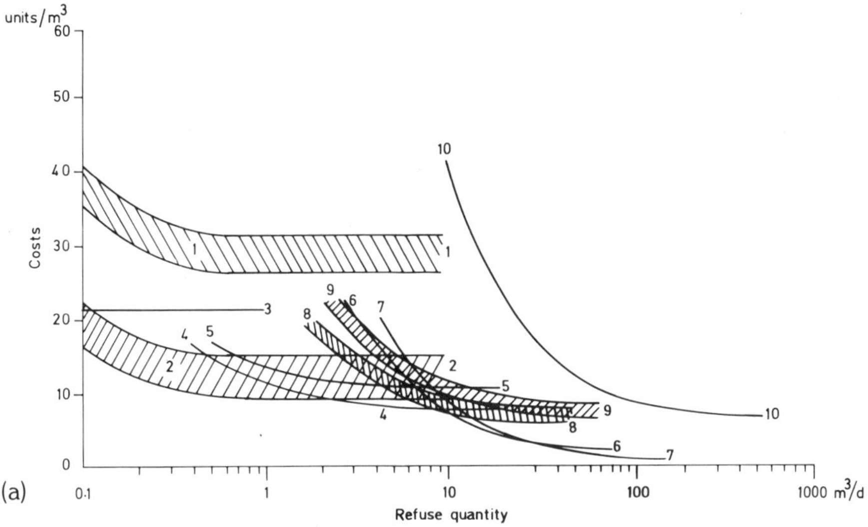

(a)

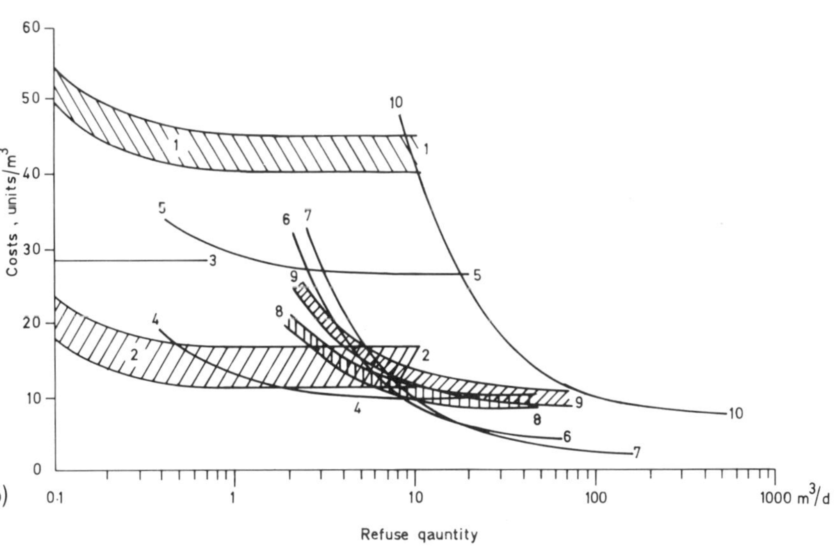

(b)

Fig. 5.12 Comparison of costs of different refuse transport systems. (a) Equipment in use 8 h/day and transport distance 10 km. (b) Equipment in use 8 h/day and transport distance 30 km.

costs and vehicle costs. Increased vehicle costs (calculated per cubic metre of loose refuse) would result from transporting less than a full load or from the extra time spent in trying to collect a full load. Account has been taken in Fig. 5.12 of the practical applications of different transport systems by calculating cost curves for amounts of refuse which would result in bins being filled on the one hand once a day and on the other once a week. The costs shown in Fig. 5.12 have been calculated for the following systems of collection and transport:

(a) 0.75-m^3 metal bins, which are exchanged for empty ones with the aid of a boom crane. A 9-tonne, two-axled truck transports eight 0.75-m^3 bins at a time, making a total load of 5.5m^3 of loose refuse. The collection time for each load is 70 min if there are two 0.75-m^3 bins at each collection point and 90 min if there is only one bin at each collection point.

(b) 0.6-m^3 reinforced plastic bins are emptied by means of a compacting refuse collection vehicle. The 16-ton, two-axled vehicle is equipped with bin-emptying apparatus. The truck can take a load of refuse from 120 0.6-m^3 bins, each bin containing an estimated 0.5 m^3 of loose refuse. The refuse is collected from a built-up area with groups of 2-6 bins at each collection point. Time-and-method studies have shown that practical values for collection times per load are 228 min if there are 8 bins at each collection point and 408 min if there are only two at each collection point[25,26]. Time losses are assumed to be included in the collection time.

(c) Refuse sacks of 0.15 m^3 are emptied on to the open platform of a two-axled light refuse collection vehicle with a total weight of 3.5 ton by means of a tipping apparatus. The capacity of the truck platform is 10 m^3. Each load comprises 80 sacks, each containing about 0.13 m^3 of loose refuse. The collection time per load is 60 min.

(d) 5.5-m^3 refuse containers are emptied by a refuse vehicle equipped with winch and compacting apparatus. The 16-ton, two-axled vehicle can accommodate 12 containers, each containing approximately 5 m^3 of loose refuse. The containers, one at each collection point, are collected from blocks and flats. The collection time for each load is 190 min.

(e) Covered refuse containers with a nominal volume of 6 m^3 and containing 5 m^3 of loose refuse are transported by a 9-ton, two-axled truck with 8-ton demountable body equipment. The collection time per container is 8 min.

(f) A 16-m^3 refuse container equipped with compacting apparatus and containing about 60 m^3 of loose refuse is transported by means of a 16-ton, two-axled truck equipped with 12-ton demountable body apparatus. The collection time per container is 20 min.

(g) A refuse container of 21 m^3 containing approximately 100 m^3 of loose refuse is transported by means of a 16-ton, two-axled truck with 12-ton demountable body equipment. The refuse is compressed by means of a separate compacting device. The collection time per container is 20 min.

(h) Seven covered refuse containers, each of 6 m^3 and containing 5 m^3 of loose refuse, are emptied by means of a 9-ton, two-axled truck with 8-ton high-lift demountable body equipment into a 35-m^3 transport container which is drawn by the same vehicle to the refuse treatment site. The refuse containers are at different collection points in areas with blocks of flats and the average distance between containers is 300 m. The transport container may be positioned at one point or moved from place to place nearer to each refuse container. This gives the extreme values for the time costs of the method. The total collection time for each transport container load is 95 min if the transport container remains at the same point all the time and 140 min if it is moved from place to place. The time required for the handling and emptying of the transport container, 12 min, is contained in the total time for each load.

(i) Ten covered 6-m^3 refuse containers, each containing 5 m^3 of loose refuse, are emptied by means of a 9-ton, two-axled truck with 8-ton demountable body equipment into the loading hopper of a 15-m^3 transport container. The refuse is compressed into the transport container by means of a hydraulic cylinder attached to the truck and compacting equipment fitted to the container. The refuse containers are situated at different points in residential areas of blocks of flats, the average distance between each refuse container being 300 m. The transport container can be positioned at one place all the time or can be moved nearer the refuse container about to be emptied, giving the extreme values for the time costs involved. The full transport container is drawn by the same vehicle to the refuse treatment site to be emptied. The total collection time for each transport container load is 150 min if the transport container remains in one place, and 210 min if it is moved. The handling and emptying time for the transport container load is 12 min.

(j) 30 covered 6-m^3 refuse containers, each with 5 m^3 of loose refuse, are emptied by means of a 9-ton, two-axled truck with 8-ton demountable body equipment into the loading hopper of a permanent transfer station. 150 m^3 of loose refuse is compressed into a 28-m^3 transport container, which is transported by a 22-ton, three-axled truck with 16-ton demountable body equipment to the refuse treatment site for emptying. The average distance from the refuse container collection points to the transfer station is 3 km. The total collection time for a full load for the 28-m^3 transport container is 560 min, in which is included 20 min for handling of the container.

The cost calculations were originally made using March 1978 purchase prices and operating costs [1,23,24]. However, recent increases in prices have not made any substantial difference to the comparative calculations. Therefore an arbitrary cost unit (unit per m^3) has been used to illustrate the level of costs in Fig. 5.12. It should be remembered, however, that the calculations made earlier were updated by means of indices reflecting Finnish conditions obtained from expert sources [22]. The cost data therefore reflect summer 1980 cost and price levels in Finland. The cost curves in Fig. 5.12 have been calculated as a function of the amount of refuse accumulated at a single collection point and are presented for practical reasons in the form of a semi-logarithmic scale. Most important calculation data are given in Tables 5.1-5.3.

Fig. 5.12 shows that in the case of small amounts of accumulated refuse it is advantageous to use a compacting refuse vehicle and appropriate bin equipment. When the amount of refuse accumulated at a collection point exceeds 1 m^3 per day, it is better to change to simple transfer load methods. If the quantity is clearly in excess of 10 m^3/day, the best solution is compacting refuse containers and compactor methods. In cases where the refuse amounts to more than 100 m^3/day, it is wise to adopt large transfer load methods. It should be noted that although curve 10 in Fig. 5.12, which illustrates a transfer load method, is higher in cost than straightforward compacting methods, there are a number of other features associated with the transfer load method, such as reduction in traffic, easier handling of refuse on reception at the treatment side, etc., which justify the choice of the transfer load method.

A comparison of Figs. 5.12a and 5.12b shows that Methods 1, 3 and 5 become considerably more expensive when the distance increases from 10 km to 30 km. It should also be noted that an increase in transport distance in other methods does not materially affect the cost level. The explanation for this is the relatively high cost of the actual collection process. The majority of the time is spent in collecting and emptying the bins and containers and only a comparatively small part of the total time is devoted to the actual transport of the refuse.

Calculating the distribution of costs using Equation 5.6, clear differences between the different collection and transport methods can be observed. Using the same principles adopted to calculate the costs shown in Fig. 5.12, the percentage costs for a number of examples are shown in Table 5.4.

TABLE 5.1 Calculation of Vehicle Costs for Compacting Refuse Collection Vehicle as an Example of Cost Calculations for Different Vehicles

	Chassis	Super-structure	Total
Purchase price (FIM)	179 000	128 000	307 000
Type price (FIM)	6 000		
Decrease of value (%/year)	30	30	

Calculation data				
Utilization degree (h/day)	4	8	12	16
Driving days (days/year)	250	250	250	250
Wage hours (h/year)	1000	2000	3000	4000
Drivers wage (FIM/h)	19.60	19.60	19.60	19.60
Personal secondary costs (%)	50	50	50	50
Fuel consumption (l/h)	8.0	7.5	7.2	7.0
Fuel price (FIM/l)	2.00	2.00	2.00	2.00
Calculation interest (%)	20	20	20	20
Depreciation time (years)				
Chassis	8	6	5	4
Superstructure	10	10	10	8
Capital cost coefficient				
Chassis	0.257	0.289	0.312	0.342
Superstructure	0.237	0.237	0.237	0.257
Fixed Costs (FIM/year)				
Capital costs	74 900	80 400	84 300	92 000
Chassis	44.500	50 000	53 900	59 100
Superstructure	30 400	30 400	30 400	32 900
Insurances	5 200	5 200	5 200	5 200
Traffic license charges	3 500	3 500	3 500	3 500
Garage and administration	5 100	5 100	5 100	5 100
Total	88 700	94 200	98 100	105 800
Variable costs (FIM/h)				
Labour costs	29.40	29.40	29.40	29.40
Tyre costs	4.10	3.40	2.70	2.70
Fuel and lubrication costs	16.00	15.00	14.40	14.00
Repair and maintenence costs	15.90	14.50	13.60	13.20
Total	65.40	62.30	60.10	59.30
Total costs				
Cost per hour (FIM/h)	154.10	109.40	92.80	85.80
Annual cost (FIM/year)	154 100	218 800	278 400	343 000

TABLE 5.2 Results of Vehicle Cost Calculations and Some Calculation Data Needed for Cost Calculations According to Equation 5.1

System no.	k_a (FIM/h)	t_k (min)	V m^3
1	79.90	70-90	5.5
2	109.40	228-408	60
3	65.30	60	10
4	117.50	190	60
5	79.10	8	5
6	100.40	20	60
7	100.40	20	100
8	84.70	95-140	35
9	102.70	150-210	50
10	139.80	540 + 20	150

TABLE 5.3 Calculation Data for Collection Bin Cost Factor k_k According to Equation 5.1

System no.	Purchase price/unit (FIM/year)	Depreciation time (years)	Calculation interest (%)	Repair cost (FIM/year)	Loss cost (FIM/year)	Annual cost (FIM/year)
1	800	3	15	30	20	400
2	650	3	15	10	15	310
3[1]	200	5	15	5	5	70
4	4 200	5	15	150	50	1 450
5	3 500	5	15	150	50	1 250
6	75 000	8	15	2 500	2 500	21 700
7	107 000	8	15	2 500	2 500	28 850
8[2]	3 500	5	15	150	50	1 250
9[3]	3 500	5	15	150	50	1 250
10[4]	377 000	10	15	7 500	7 500	90 100

[1] Sack holder only. Sack price 1.30 Finnish marks is taken into account when total bin costs are calculated.
[2] One container only. Transport container costs are included in vehicle costs. Collection container cost corresponding to 1/7 of refuse quantity in transport container
[3] One container cost only. Transport container costs are included in vehicle costs. Collection container costs corresponding to 1/10 of refuse quantity in transport container.
[4] Transfer station costs only. Collection container costs are taken from system 5, however, corresponding to 1/30 of refuse quantity in transport container at the transfer station.

Pipeline Methods

Comparing the costs of pipeline systems of refuse collection and transport with traditional methods poses a considerable problem. This is because of the vast differences between the two types of system, for example in the first transport stage from the place where the refuse originates. In general, however, it can be

TABLE 5.4 The Structure of Refuse Collection and Transport Costs Illustrated by Some Examples

Method	Example: Transport distance (km)	Example: Quantity of refuse (m³/day)	Indirect transport costs (%)	Actual transport costs (%)	Bin and compactor costs (%)
	(Total Collection time 408 min)				
Refuse bin/ compacting refuse vehicle	10	0.1	60.2	5.1	34.7
	10	0.5	83.4	7.0	9.6
	20	0.5	79.2	11.7	9.1
Refuse container/ truck with demountable body equipment	10	1	12.7	54.3	33.0
	10	5	17.2	73.8	9.0
	20	5	11.1	83.1	5.8
Compacting refuse container/truck with demountable equipment	10	10	7.2	12.2	80.6
	20	30	12.9	38.7	48.4
	30	60	13.5	61.0	25.5
Compactor/refuse container/truck with demountable body equipment	10	50	15.1	25.9	59.0
	20	50	12.7	37.9	49.4
	40	100	12.9	61.9	25.2

said that pipeline methods are capital intensive by reason of the construction costs involved, while in the case of traditional methods of collecting and transporting refuse the part played by labour costs is considerable, and may even be decisive. With pipeline methods the proportion of fixed costs may be of the order of 75% of the total costs[10]; consequently, increases in the general cost level, resulting from inflation, increased wages, etc., have only a small effect on the total costs of pipeline systems of refuse collection and transport.

The total costs of pipeline systems are usually significantly higher during the first years that they are in use than the corresponding costs for traditional collection and transport methods because of the high investment cost. Manufacturers' calculations for pipeline systems usually cover 5-10 years, and it is only after this period that pipeline methods of refuse handling become more economical compared with conventional methods. Fig. 5.13 illustrates a calculation which indicates that after 6-7 years the suction pipeline method of refuse handling is a very economical method[28].

A comparison has been calculated for two different residential areas. In the first area (A) there are 1200 homes in blocks of flats. These are connected to a suction pipeline system with an average of 30 homes to each connecting valve. The total length of the transport pipeline is 1200 m, and the suction machinery is dimensioned to cope with 300 m^3/min. In the other area (B) there are 3000 homes, some of them in blocks of flats and others in terrace houses. The average number of homes to each connecting valve of the suction network in this case is 12. The total length of the transport pipeline is 6000 m, and the machinery of the network has an effect of 1400 m^3/min. The emptying of refuse

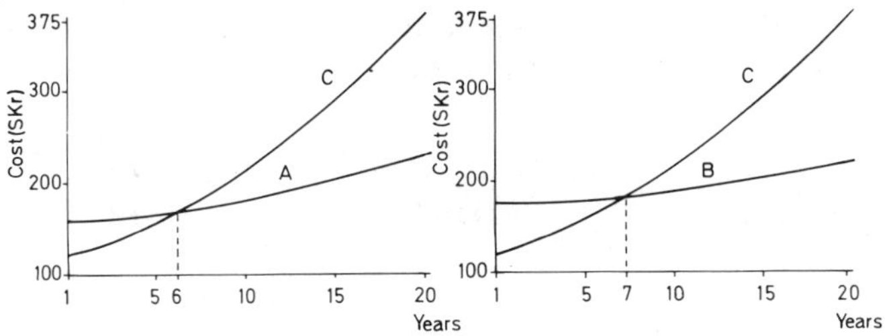

Fig. 5.13 Comparison of costs for vacuum pipeline systems and traditional methods of transporting refuse[28]: A: vacuum pipeline installed in blocks of flats. B: vacuum pipeline installed in flats and one-family houses. C: traditional methods.

is carried out twice daily. In both cases the investment costs for the system are written off over a period of 30 years and an annual interest rate of 8% is assumed. The quantity of refuse accumulated is 25 l/day/home, which corresponds to normal Swedish amounts. The quantity is assumed to increase by 3%/year. The annual rate of inflation is assumed to be 6%.

The costs for such a suction pipeline method are compared with the costs of traditional methods of refuse collection and transport which, at the time the calculation was made, amounted to 120 Swedish kronor per home per week. Refuse was emptied twice to three times a week.

Table 5.5. shows the details of the calculated annual costs of refuse disposal per home during the first year of use. Certain economies are achieved with

TABLE 5.5 Comparative Costs of Suction Pipeline and Traditional Methods of Refuse Collection and Transport

Type of cost	Costs of different methods (SKr)		
	A	B	C
Investment costs			
Machinery and construction work	1000	1100	
Pipeline conveyor, electric wiring, compressed air lines and chutes	360	1800	
Transport air and connecting valves	340	2200	
Total investment costs	1700	5100	
Annual capital costs (30 years 8%)	151	453	
Annual operating costs			
Electricity and maintenance	30	40	
Transport of refuse containers	10	25	
Total annual operating costs	40	65	
Total annual costs	191	518	
Total annual costs per home during the first year (SKr/year)	159	173	120

pipeline methods of waste disposal since less space is required for refuse handling facilities and also roads can be dimensioned for less traffic. It should be noted, however, that other service functions make a road network necessary and this very often has to be of such a size as to permit truck traffic (removal vans, etc.).

PROSPECTS FOR THE FUTURE OF WASTE DISPOSAL

It is to be expected that the significance of waste disposal as one form of communal service will become clearer during the 1980's. A tempering of the over-exaggerated stress on environmental conservation on the one hand and a less disparaging attitude towards waste disposal problems on the other hand will make it possible for local authorities to put practicable refuse handling systems into effect at reasonable cost. However, it should be noted that it is the service and efficiency of waste disposal systems which dictate to a large extent the costs of such systems.

Labour safety and hygienic factors are likely to carry increasing emphasis in the future. This will mean that waste disposal will have to be organised in such a way that it does not entail human handling at all and also that employees engaged in waste disposal are not subjected to health risks at any stage of their work. Labour safety legislation will become more strict and as a result refuse collection will become more mechanized. A possible alternative to this would be to transfer part of the responsibility for waste disposal to the ordinary citizen; however, such a prospect is hardly probable. In densely populated areas and in large blocks of flats, the use of pipeline methods of refuse handling is likely to become more widespread. Aesthetic considerations will result in the disappearance of refuse from yards and roadsides. On the other hand, labour safety factors will make necessary a more careful planning of waste disposal facilities so that refuse collection areas in residential areas are, from an aesthetic, hygienic and labour safety point of view, absolutely impeccable.

An important factor affecting refuse collection will be the development of the pattern of residential planning. If the tendency towards small, one-family houses continues, the mechanized collecting and emptying of refuse bins, if these continue to be used, will become more common. If, on the other hand, there is a move back to living in blocks of flats because of the energy situation, it is likely that we shall see more pipeline systems in use.

The transport of refuse is greatly affected by the choice of refuse handling system. In the future the recycling of waste and the accompanying sorting processes may be expected to be of particular significance. Because of the sorting special vehicles equipped with different compartments for different kinds of refuse will need to be developed. However, the prospects as far as advanced sorting of waste at its place of origin are concerned do not seem too bright at the moment. Nonetheless, it is probable that presorting of the type now taking place, e.g. the separation of paper and rags from other household refuse, will be extended to cover other types of refuse although this will probably not go too far. This will mean that refuse collection and transport equipment will have to be adapted to the type of refuse system in use. It is likely that collection of refuse will be done by means of light and relatively simple vehicles while the transport of it will be carried out by efficient vehicles especially suited to this purpose. Instead of developing complex special vehicles, it will be necessary to develop simple multipurpose vehicles.

The need for transfer loading will depend largely on the development of refuse treatment methods. If the treatment of refuse is to be concentrated to large centres which, by reason of their demands on space, will have to be situated outside towns and cities, then the need for transfer loading will no doubt increase. If, on the other hand, the treatment of waste is spread over many small units which can be situated within the normal town planning area, there will hardly be any need for transfer loading. The exploitation of refuse for making fuels and humus will favour the siting of refuse treatment plants near centres of population, in which case the need for large transfer stations will be significantly reduced. Although the need for transfer loading can be expected to grow in the near future, in the long run transfer loading may perhaps not play a very important role.

Developments in refuse treatment are likely to lead in the first instance to the evolution of ways of producing fuels that can be burnt in those boilers at present in use with causing any ill effects. The part played by the conventional incineration of refuse will probably decline into insignificance. The need for refuse tips is likely to be considerable in the near future, since it will be necessary to find suitable places for the storage of difficult wastes. Moreover, it is unlikely that refuse treatment methods will be developed to such a degree that all products of refuse treatment plants can be recycled or used for landfill purposes, so reducing the need for refuse tips. The need for humus products will in all probability swell with the increase in green areas in towns. As a result there will be an increase in the use of biological methods and composting processes for the treatment of refuse. In the long run, increasing account of ecological restrictions will be taken when considering ways of treating refuse. Consequently, ecological factors and, in the short run, economic factors will play a larger role than they do at present.

Another long-term prospect is that the total energy needs in refuse handling systems will have to be minimized in the same way that the energy needed for other activities will have to be reduced. This will mean that studies of ways in which to conserve energy will extend beyond the traditional field covered by refuse handling in order to achieve the maximum minimization of energy needs. The coordination of materials and energy will be subjected to more careful scrutiny than hitherto.

When planning waste disposal in the future it is to be expected that more attention will be directed to all levels of planning and all time intervals. Furthermore, the planning of products and packaging will perforce take more account of what happens to the product or packaging when it has been used. In the first instance this will affect disposable articles but in the long term consumer durables will also be subjected to considerations of this kind. All kinds of waste disposal are likely to be characterized by increasing concern on the part of society at both the national and the international level in the near future. With the aid of a practical attitude towards refuse handling problems it will be possible to satisfy the justified demands of future generations for clean air, clean water, a thriving countryside and an environment not polluted by noise.

REFERENCES

1. PAATERO, J., 'Waste Management Systems', Part I: *Suomen Kunnat* 7 (1979) 483; Part II: *Suomen Kunnat* 8 (1979) 595 in Finnish.
2. Illegal refuse rooms, kurskompendium, Bygginfo, Stockholm (1979) in Swedish.
3. KERSTAN, H.-G., 'Das neue Umleersystem MGB 5.0', *Müll und Abfall* 8 (1978) 249.

4. DEWEY, W.-J., GLADBACH, D. and NEUHAUS, N., 'Wirtschaftlichkeitsberechnung in der Abfallbeseitigung für die Bereiche Sammlung und Transport', *Müll und Abfall* 8 (1978) 251.
5. MÄKINEN, K., 'Refuse Transfer Loading', *Helsingin kaupungin kaupunkisuunnitteluviraston yleiskaavaosaston julkaisuja YB:10/79*, teknis-taloudellinen toimisto, Helsinki (1979) in Finnish.
6. MILLARD, R.F., 'The Handling and Transportation of London's Waste — Design and Operation of Transfer Stations', *Proceedings of the 1980 ISWA Congress*, London, 1980.
7. OSCHLIES, D., 'Plan and Operation of Transfer Stations in Berlin', *Proceedings of the 1980 ISWA Congress*, London, 1980.
8. BROWN, M., VENCE, T. and REILLY, T., 'Transfer Stations Offer Time, Energy, Cost Savings in Variety of Situations', *Solid Wastes Management* 22 (12) (1979) 28.
9. POLLOCK, E.L., 'Underground Transfer Station Carved Out of Rock in Downtown Stockholm', *Solid Wastes Management* 22 (1) (1979) 66.
10. LUITEN, P.H., 'Vacuum Systems for the Collection of Solid Wastes', *Proceedings of 1976 National Waste Processing Conference held in Boston, May 23-26*, 1976, ASME, New York, 1976.
11. LUITEN, P.H., 'Pneumatic Refuse Collection — a New Utility,' *Consulting Engineer*, April 1976, 76.
12. JAEHN, M., 'Planung und Bau sowie Wirtschaftlichkeit von Müllsauganlagen', *Heizung-Lüftung-Haustechnik* 26 (1975) 145.
13. JAEHN, M., 'Erfahrungen und Tendenzen beim Bau und Betrieb von Müllsauganlagen', *Heizung-Lüftung-Haustechnik* 28 (1977) 21.
14. MÜLLER, H.-J., 'Abfallsammlung und -transport', *Kommunalwirtschaft* 9 (1976) 292.
15. ZETTERSTRÖM, R., 'The Centralsug System for Automatic Refuse Collection', *ISWA Information Bulletin No. 11/12*, December 1973, 82.
16. WEEDA, H.W.J., 'Pipeline Systems as Non-conventional Systems of Waste Storage and Collection', *Proceedings of the 1980 ISWA Congress*, London, 1980.
17. BODELIUS, L. 'Comments on Pipeline Systems as Non-conventional Systems of Waste Storage and Collection', *Proceedings of the 1980 ISWA Congress*, London, 1980.
18. SAARENTO, H., 'Transportation and Contract Work of Municipal Refuse', *Maarakennus ja kuljetus* 9, (1978) 356, in Finnish.
19. RINNE, J. and HAAPANEN, M., 'Transportation Systems', Ekondata Oy, Helsinki (1978) 173, in Finnish.
20. SIBIGA, J., 'Analyse der Transportfunktion bei der Müllabfuhr', *Müll und Abfall* 5, (1976) 152.
21. TABASARAN, O., 'Einsammlung und Transport von Müll mit Front- und Hinterkipperfahrzeugen und 4.4 m^3-Grossbehältern', *Müll und Abfall* 8, (1976) 234.
22. HAAPANEN, M., Personal communication, 1978.
23. PAATERO, J., 'On the Technology of the Collection and Transportation of Municipal Refuse in Finland', paper presented at the environmental symposium held 29-31 May, 1978, in Helsinki, in Finnish and in Russian.
24. PAATERO, J., 'Transportations Related to Municipal Waste Management', paper presented at the waste handling symposium held 26-28 April, 1978, in Piwniczna, in Finnish and in Russian.
25. AHTO, I. and HORKKA, J., 'Transportation Costs in Municipal Refuse in

Helsinki', thesis at Helsingin Kauppiaiden Kauppaoppilaitos, Helsinki (1974) in Finnish.
26. PERTTULA, P., 'Transportation Economy of Waste Management', SITRA's YVY-project E20, Helsinki (1975) in Finnish.
27. PAATERO, J., 'Characteristics of Different Waste Management Systems at an Alteration Period', paper presented at the waste management symposium held 13-14 December, 1979, at Hyvinkää, in Finnish.
28. Refuse Transport in Residential Areas, informationsbroschyr, Centralsug Ab, in Swedish.

Dr. Paatero has wide experience in solid waste management and sewage sludge handling, having acted as a scientist, R&D engineer and consultant both in Finland and abroad (including the Middle East and CMEA countries). He has delivered numerous articles and lectures in several countries, being one of the few Finnish waste management experts, and is an expert member in the Waste Management Advisory Council in Finland and in several Finnish state organizations. He is now responsible for the long-range development of environmental engineering by the Development Centre of Oy Partek Ab.

6 Source or Central Separation of Household Food Waste

Jens Chr. Tjell, Anton Elmlund and Jens Aage Hansen

INTRODUCTION AND BACKGROUND

Utilization of waste food could be regarded a rather trivial issue: feed stock, pets and other animals have always had leftovers from man's meals. Organized salvage is another well known practice, particular to urbanized settlements where considerable amounts of stock food can be collected, if necessary, and when adequately organized. War times have certainly proved both the necessity and the feasibility of such salvage schemes; e.g. in England during World War II. Collection and re-use of waste food as feeding stuff has diminished since the war[1]. The reasons for this seem to be ample supplies of raw materials and energy, and hence cheap industrially manufactured stock feed. Also, in the same period, labour-intensive collection and separation systems were systematically ruled out by shortage of labour. And finally, disposal systems were offered that would accept any mix of waste materials from households, restaurants, offices, warehouses, department stores, etc.

However, shortage in supply of certain raw materials and energy are very obvious challenges to the society of today and the future, so recycling has become the new key word for organized salvage. And very recently, at least in Norway[2] and Denmark[3], recycling of food waste for stock feed has become a subject of much interest.

When entering upon new recycling schemes, health considerations should be of primary concern. To implement or even improve health in a society where increased recycling is set as a goal involves an inherent problem of augmented epidemiological risk, in terms of contagions, toxic substances and other hazardous compounds. Also an economic assessment and a practical feasibility study should accompany any new recycling proposal before implementation. Energy requirements will of course play an important role in this respect.

FOOD WASTE SALVAGE SYSTEMS

Food waste from households, hospitals, restaurants and nursing homes can be separated from other waste flows and re-used if found adequate.

This chapter will not advocate a particular technically feasible route for re-use of food waste; however, to justify the subsequent considerations on quality and quantity a conceptual design of a food waste recovery system is outlined in Fig. 6.1.

Several food waste recovery systems do currently exist[1,3], which gives this exercise a realistic background. Unlike the on-going experiments and full-scale operations, one possibility is that of feeding fur animals instead of pigs or other

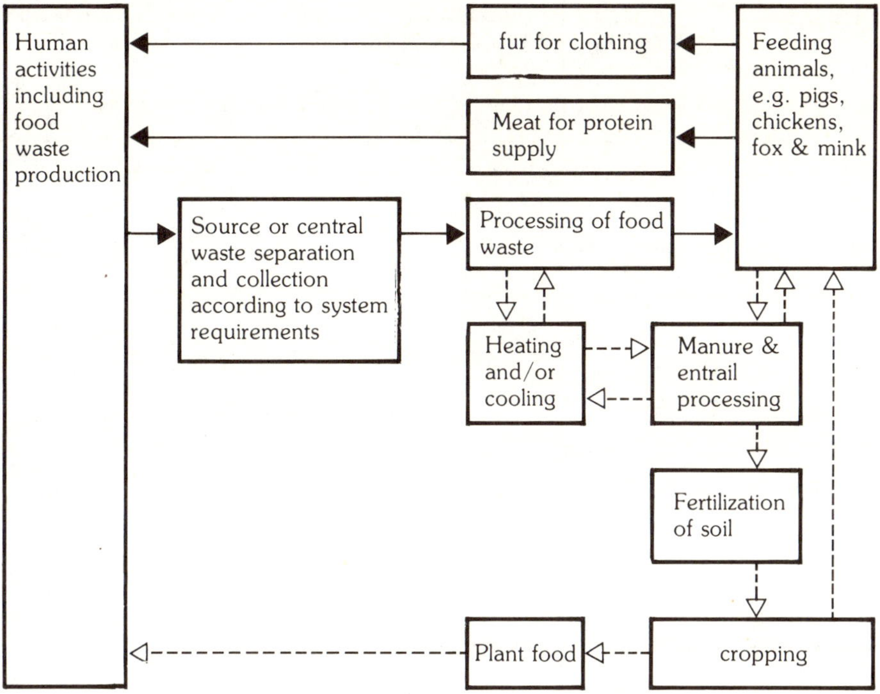

Fig. 6.1 Sample system approach to food waste utilization and management. Solid lines express the major flows in terms of energy and economy.

animals where the meat is normally used for human consumption. Reasons for this would be limited risk of infectious deseases, less demand for infection control and hence cheaper processing, relatively small amounts of fodder based on food waste compared to total fodder consumption, and convenience to man using fur coats, particularly in temperate climates. Possibly the fox would be an ideal consumer of salvaged food waste, cf. Fig. 6.1.

CHARACTERIZATION OF URBAN FOOD WASTE

Danish household waste composition has not changed much in food waste content over the last 80 years, although other main components have changed dramatically. This is illustrated in Table 6.1, where typical modern refuse from the Copenhagen area is compared to refuse from the same area around the turn of the century. The total quantities of per capita collected refuse are nearly identical, and the food waste (animal and vegetable products) constitutes in both cases approximately one third, presently rated at 80 kg wet stuff per capita per year.

The most pronounced change in composition stems from the lack of ashes and the abundance of paper in modern refuse. Amounts of polymers, glass and iron have increased, obviously as a consequence of the much higher packaging materials turnover in the present consumer society.

Turning to the amounts of potentially recoverable food waste in Denmark,

TABLE 6.1 Typical Composition and Amount of Danish Household Waste in 1979 and in 1902–03

Component	Average wet weight distribution (%) 1979[3]	1902–03[4]
Paper	27.3	9.5
Cardboard	7.4	—
Animal products including bones	1.8	33.5
Vegetable products	33.2	
Textiles	1.9	2.5
Rubber, leather	0.4	—
Plastics	5.9	—
Other combustibles including wood	5.6	1.0
Clean glass	4.0	0.8
Coloured glass	3.8	0.9
Scrap iron	2.4	1.4
Other metals	2.0	0.1
Other non-combustibles, coke, coal, ashes, slag and porcelain	4.3	50.3
Total (%)	100	100
kg (wet)/year/person	221	180

Table 6.2 indicates a total of 90 000 ton dry matter yearly. Of this, only approximately 50% can realistically be recovered through practical methods of separation and collection, although the rating for large institutions as hospitals may be as high as 100%.

Compared to the yearly consumption of fodder in Denmark, i.e. 13 x 10^6 ton dry matter, the recoverable fraction of food waste constitutes only 5%, and in monetary terms approximately 100 x 10^6 Dkr. The most pertinent comment on this statistical fact is that extensive recovery may give rise to a good deal of employment, but have a rather marginal effect on the national economy. Although the food waste potential in households is the highest, it also constitutes a real problem economically to collect these comparatively small and dispersed contributions.

Table 6.3 summarizes experiences in Denmark, Norway and Italy as to the composition and nutritive quality of food waste recovered by different separation systems. Generally speaking, food waste proper is similar to vegetable matter such as beet, except for a rather high fat content. Most of the food waste is classified as vegetable matter (Table 6.1), but obviously fat of animal origin is disposed of in rather large quantities.

When food waste is recovered centrally by mechanical means, two distinctive changes occur: the fibre and the ash content increase. Obviously this can only be ascribed to increased, efficient mixing of the food waste proper, with glass, iron and paper particles intimately adhering to the wet and sticky food waste

TABLE 6.2 Potential Danish Recovery of Organics for Stock Feeding According to Household Participation, Collecting System and Separation Techniques

Source of feedstuff and assumptions as to separation and collection techniques	Speculative total per year (ton dry matter)
1. Households, 80 kg, wet/person/year, source separation and collection, 100% participation, 20% dry matter*	82 000
2. Households, 60% participation in (1)	49 000
3. Households, collection of mixed waste for central disintegration and separation, max. available, estimated 30% recovery (pilot scale observations)	24 000
4. Additional source collection of food waste proper from	
hospitals (0.5 kg, wet/bed/day°)	2500
nursing homes (0.6 kg, wet/occupant/day§)	5000

*5.1 x 10^6 citizens in Denmark.
°approx. 43 000 beds in Denmark.
§approx. capacity equals 48 000 occupants.

TABLE 6.3 Nutritional Characterization of Urban Waste-Derived Fodder Compared with Some Typical Danish Stock Feeding Crops

Feedstuff source and information reference	Dry matter content (%)	Crude protein %*,6	Fat %*,6	Fibre %*,6	Ashes %*,6	Carbohydrates %*,6
Barley grain[5]	86	12.6	2.2	4.8	2.5	77.9
Beet[5]	16	7.4	0	6.2	6.8	79.6
Beet top[5]	20	16.4	2.0	9.5	19.7	44.5
Institutional food waste including hospitals, nursing homes and restaurants — Norway[2]	22	22.4	22.5	2.9	8.5	43.7
Institutional food waste including hospitals and nursing homes only — Denmark[7]	missing	27.1	21.5	4.5	5.8	41.4
Household food waste, source separation — Norway[8]	17	20.0	20.0	5.3	10.6	44.1
Household waste organics, all vibration/sieving fractions combined, net weight average§[9]	missing	16.8	15.9	10.7	22.3	34.3
Household waste, central Rome, mechanical separation, analyses in Denmark[10]	missing	12.6	11.4	38.0	16.2	21.8

*All % expressed on a dry matter basis.
§Adjustment for variation of % water still to be done.

particles. This may eventually be a determining factor in selection of viable recovery systems, because composition and quantity of recovered organics are changed significantly.

TABLE 6.4 Heavy Metal Contamination of Waste-Derived Fodder Compared with Some Typical Danish Stock Feeding Crops

Feedstuff source and information reference	Metal (ppm dry matter basis)				
	Lead Pb	Mercury Hg	Cadmium Cd	Zinc Zn	Copper Cu
Barley grain*	0.15	0.001	0.04	26	3.1
Beet*	1.0	0.003	0.30	27	4.2
Beet top*	5	0.07	0.75	47	12.7
Institutional food waste, PNA[7] Denmark	0.33	0.1	0.10	119	—
Household food waste separated at source and collected separately, Gendan[3]	0.72	0.1	0.16	29	9.3
Household food waste mixed at the source and separated manually at a collection station, Gendan[3]	11.7	0.1	0.45	119	18.4
Mixed household wastes separated centrally by vibration sifters°					
all fractions, organic mix, MP[9]	96	0.12	1.4	416	40.1
>30mm, organic mix, MP[9]	28	0.01	0.7	155	14.2
Mixed household wastes separated mechanically at ESMIL, Holland, analyses by Muus[10]	133	1.2	1.6	517	53
Mixed household wastes separated mechanically at the Cecchini plant in Rome; fodder pills analyzed					
Muus[10]	62	2.1	3.20	380	82
Kvärner[14]	69–70	0.2–1.1	1.6–1.7	272	68
Danish fodder legislation, max.	5	0.1	—	250	125§

*Analyses for Pb and Cd by Tjell[11]; for Hg by Jensen and Sevel[12]; for Zn and Cu from Min[13].
°Adjustments for varying water content may change the figures slightly.
§Depending on animal species.

The separation method employed exerts even stronger influence on the heavy metal content in recovered food waste. As with main constituents, food waste proper is very much like vegetable matter with respect to the metal content observed in different investigations (Table 6.4). The increases in metal content in food waste observed after central hand-sorting of refuse are rather large, but mechanical sorting produces by far the largest increases. Much speculation has been devoted to this phenomenon, and one possible explanation is presented below. Whatever the origin of the metals, the resulting concentrations of the essential elements Zn and Cu are high, although the regulatory limits for fodder are not seriously violated. For the toxic metals lead and mercury, the resulting concentrations in mechanically sorted food waste are high and exceed the regulatory limits excessively.

HOUSE DUST AS A SOURCE OF TRACE METAL CONTAMINATION OF ORGANICS IN HOUSEHOLD WASTE

The trace metal content of organics recovered centrally and shown in Table 6.4 are well above the natural trace metal contents found in food waste separated at source in the kitchen. The observed contents of lead and cadmium would be prohibitively high for usage of the recovered organics as fodder, especially to pigs meant for human consumption.

The increase in metal content of waste organics is moderate when recovered manually from refuse sacks. It is, however, quite substantial when organic material including food waste is recovered through a mechanical process, e.g. involving sifting as the main separation principle.

In an attempt to identify possible sources of metals in household waste, dust samples from 18 homes were collected and analyzed. The average amount collected from vacuum cleaners was 1.2 kg dry matter/person/year, and the average contents of four metals are shown in Table 6.5

If it is assumed that 50% of the dust adheres to the wet food waste particles, it is possible to account fully for the observed increases of lead, cadmium and copper in the centrally hand-sorted food waste. The observed increase in zinc content is twice the calculated.

The observed increases in concentration of all four metals in the mechanically sorted organic waste cannot be explained by this dust hypothesis, but the so-called organic fractions contain several per cent of metals (presumably iron), which according to several investigations carry appreciable amounts of the toxic metals lead and cadmium from soldering and pigments.

Apparently, a significant part of the trace metal contamination of recovered waste organics can be traced back to contamination by house dust disposed of in the refuse sacks.

SOURCE OR CENTRAL SEPARATION

Separation of food waste from households and institutions (hospitals, nursing homes, restaurants, etc.) deserves attention for several reasons. For instance, the separation of such organics might influence significantly the type and frequency of collection of other waste components; handling, treatment and recycling of other waste components, e.g. paper; design of many intermediate steps in a processing line; and last but not least the operation of the sanitary landfill, where both leachate and gas generation and control could be influenced dramatically. However, focal points in this chapter will be limited to fodder production (recovery of food waste organics), contamination and economy, with special emphasis on source versus central separation methodologies.

TABLE 6.5 Observations and Speculations on Heavy Metals in Food Waste and Organics from separated mix wastes

House dust estimate— 1.2 kg dry matter/person/year cf. Tjell[15].
Food waste estimate — 80 kg wet weight/person/year — 16 kg dry matter, cf. Gendan[3].

Item	Concentration (p.p.m. dry matter)			
	Lead Pb	Cadmium Cd	Zinc Zn	Copper Cu
Basics				
Households' own separation Gendan[3]	0.72	0.16	29	9.3
Manually sorted mixed household waste, Gendan[3]	11.7	0.45	119	18.4
Mechanical, central sorting at ESMIL, Muus[10]	113	1.6	517	53
House dust examination average, Tjell[15]	295	7.0	1260	390
Increases observed				
Food waste proper augmented to manually sorted mix, item 2 less 1	11	0.3	90	9
Food waste proper augmented to mechanically sorted mix, item 3 less 1	132	1.4	488	44
Increases calculated				
Food waste proper + 3.6% dust corresponds to 50% of generated dust	11	0.25	46	14
Food waste proper + 10% dust corresponds to 150% of generated dust	29	0.7	127	39

Fodder from Food Waste

The potential fodder amounts available in Denmark from recovery of food waste would amount to approx. 90 000 ton/year or approx. 5% of the present fodder consumption. Obviously, only for local and possibly specialized applications would the food waste-derived fodder deserve attention, cf. the section on food waste salvage systems above.

For the food waste-derived fodder, another complication has been emphasized; namely that of quantity and quality, which are linked concepts. If, for instance, a limit on the presence of a certain contaminating agent is imposed on the fodder before use for pig feed, this limit may imply that only a few of several technically feasible handling schemes could be applied; and/or that only a certain fraction of the recovered food organics could be utilized, given the handling and processing methodology. Quantity and quality are not concepts that can be dealt with separately, and this has a direct relation to the choice between source and central separation of household and similar wastes.

Contamination

Contamination of salvaged food waste could be either inherent to the waste product or a result of handling and processing. Only the latter is dealt with here, and only heavy metals are used as examples of persistent trace contaminants.

As shown above, the risk of contamination of organics recovered from household waste is real. And mixing of several waste products, such as food leftovers and house dust, seems a most important and possibly the critical operation.

The least possible contamination is incurred when the food waste is kept separate from the moment of orgination to the time of feeding animals. Also the potentially largest amount of food waste can be salvaged in this way. However, this may imply both inconvenience to the household and increased cost of collection, transportation, processing and distribution of salvaged organics.

A most thorough contamination of salvage organics is incurred when all waste products are mixed in the household and then separated centrally after homogenization by some mechanical device, cf. sections above. Depending on fodder quality criteria, the amount of food waste salvaged may vary, but indications are that the maximum recovery is below 30% (by weight) of the potentially available food waste. Such technical solution may seem more convenient and hence more attractive to the individual household.

The choice of separation methodology (central or source) would, however, not normally be made only on premises regarding contamination. Other considerations would be resource recovery, environmental protection, occupational health, employment and economy.

Economy of Source Separation Systems

Table 6.6 elucidates some economical aspects of source separation. Five different approaches have been delineated: the basic technical assumptions are shown in the notes to the table, and price indications are given in each of the five columns. An attempt has been made to obtain realistic figures pertinent to a case involving an existing Danish city community of 42 000 people.

A range of fodder prices appears in Table 6.6. As can be seen, prices may vary by a factor of four, depending on the kind of collection and processing system offered to the waste producer (household or institution). The prices given, varying from 1400 to 5800 Dkr/ton dry matter could be interpreted only very cautiously in absolute terms. The relative variation shows, however, that the price of fodder produced from food waste would be extremely sensitive to choice of collection and processing procedure, which again is linked to fodder quality criteria, cf. the section on contamination, above.

It is interesting that only for the alternative where collection costs to the fodder producer are reduced to a bare minimum would the price of food waste-derived fodder compete realistically with the price of virgin products such as grain from barley. This indicates the need for an in-depth analysis of the total salvage/disposal system before decisions are made. Such analysis may indicate the advantages of a source separation system, not necessarily on the grounds of big money in fodder from food waste, but rather related to facts on the closer control of waste flows in society and environment together with a reasonable economy and an improved resource conservation programme.

TABLE 6.6 Cost Range Estimates, Danish Kroner per ton Dry Matter, for Production of Fodder from Food Waste Based on a 42 000 capita Community in Denmark

The case refers to the city of Vejle, Denmark. The assumptions made are: 42 000 capita equivalent to 18 000 households. Households produce approx. 3360 t DM/y and institutions (hospitals, cafeterias, inns, hotels) approx. 370 t DM/y of food waste. Collection efficiency is 100% for institutions, 50% for households. Use of fodder is at the processing site. Data provided by Gendan, Denmark. Reference: 1 ton dry matter food waste is equivalent to 1.2 ton barley at price of approx. 1600 Dkr.

Item of expenditure to the fodder producer	Technical alternative: Institutions and households			Institutions only	
	High[a]	Medium[b]	Low[c]	High[d]	Low[e]
In-house collection					
depreciation	207	207	207	0	0
wages & salaries	0	0	0	0	0
operation & maintenance	1480	1037	0	0	0
Collection and transportation					
separate collection					
depreciation	283	49	49	270	270
wages & salaries	2195	122	122	1351	675
operation & maintenance	261	63	63	284	351
combined collection					
depreciation	0	49	49	0	0
wages & salaries	0	976	0	0	0
operation & maintenance	0	0	0	0	0
Processing					
depreciation	673	538	538	1009	807
wages & salaries	244	244	244	40	40
operation & maintenance					
pasteurization	161	161	161	161	161
drying	310	0	0	310	0
Total: Danish kr./tons dry matter	5814	3426	1433	3405	2304

[a]This alternative implies separate collection of food waste for both households and institutions once a week; for households 8 collectors and 4 trucks are used, for institutions 1 collector and 1 truck; plastic bags and bag containers are paid for; one labourer operates the processing which includes homogenization, pasteurization and drying.

[b]This alternative implies: combined collection once a week of food and other waste from households using a specially equipped truck; household food waste is thus collected at the price of one extra labourer for the truck crew; separate collection twice a month from institutions (this collection is paid for); plastic bags (cheapest type) and bag containers are paid for; one labourer attends the processing plant, which includes homogenization and pasteurization but not drying, i.e. wet fodder is produced.

[c]This alternative resembles (b) except that the fodder producer receives all household food waste free of any charge (nothing for bags, nothing for collection labourer).

[d]This alternative implies: institutions provide their own bags and bag containers; separate collection is once a week by 1 truck and 1 labourer; processing includes homogenization, pasteurization and drying in a batchwise operation amounting to 1 h/day for equipment and 1 labourer; dry fodder.

[e]This alternative resembles (d) except that collection is twice a month and drying is excluded; wet fodder.

The apparently troublesome economy of a source separation/food waste recovery programme cannot be used as justification for a central sorting system. Firstly, in the two instances the fodder output is different, both quantitively and qualitatively. Secondly, two overall waste-handling systems based on either source or central separation as the key elements would come out entirely different on implications as to environmental pollution from landfills, economy of the total disposal system, and which future disposal and recovery systems are feasible, technically and politically.

CONCLUSIONS

Heavy metal concentrations for food waste salvaged after source or central separation are presented, with some cost data to demonstrate possible price variations for waste-derived fodder, assuming different collection and processing systems applied to a case city with a population of 42 000. Based on these facts and speculations some conclusions are proposed with special relation to food waste recovery.

(i) *In favour of source separation.* The lowest possible contamination is achieved, i.e.. the best quality of fodder. Potential recovery is also the best in terms of quantity. Recovery of other products, e.g. paper, is facilitated. Recovery of heat from the rest of the household waste is enhanced. Gas, leachate, bird, insect and other controls at the landfill site may be more easily accomplished where landfills are necessary.

(ii) *Against source separation.* The inconvenience to the household may be noticeable. The cost for separate collection may exceed that of mixed waste collection.

(iii) *In favour of central separation.* The convenience to the household may be maintained at the same time as recovery takes place centrally — to a certain extent. The collection cost may be kept at a minimum.

(iv) *Against central separation.* Food waste recovery will be more difficult the more severe the fodder quality criteria. In general, the degree of recovery of each waste component (e.g. food, paper, energy) is reduced. Future recovery operations in waste management may be excluded because high investments in central separation plants tend to commit decision makers, both public and private.

Other waste management alternatives exist than those indicated above. Therefore, to assess a given problem adequately, e.g. for a country, a municipality or a city, a more comprehensive analysis would be necessary. Systems analysis would be a tool of great use for any such case.

Thus, in short, source separation seems to imply the greater potential for efficient recovery of materials and products from the household waste flow. However, to implement source separation some extra household participation seems necessary, and possibly the benefit to the household must be found through improved common weal, not through private economic compensation.

REFERENCES

1. MAFF (Ministry of Agriculture, Fisheries and Food) Her Majesty's Stationery Office, London 1955.

2. HEIE, A.C., NATVIG, K., HALMO, T.M. and KNAP, A.H., *Application of Previously Produced Data on Food Waste in Notudden Municipality* (in Norwegian), Department of Technical Chemistry, University of Trondheim, Norway, March 1976.
3. GENDAN, *Waste Flows through Private Households in Denmark*, (in Danish), Report Series, Nos. 2 (December 79) and 4 (May 1980), St. Kongensgade 27, DK-1264, Copenhagen, Denmark.
4. CPH, *Refuse in Copenhagen* (in Danish), published by J.H. Schultz, 1904, Copenhagen, Denmark.
5. ANDERSEN, P.E. and JUST, A., *Tables on Composition of Fodder* (in Danish), Det kgl. danske Landhusholdningsselskab, Landhusholdningsselskabets Forlag (Odense), Copenhagen 1975.
6. AM, *Working Methods for Chemical Analysis of Milk, Dairy Products and Fodder*, published by J.H. Schultz, Copenhagen, 1958.
7. PNA, *Project New Work* (in Danish), Progress report as of Oct. 1, 1979; distributed by Erik Bendixen, Langelinie 2–4, DK-6000 Kolding, Denmark.
8. LYSO, A., *Food Waste as Fodder for Swine* (in Norwegian), paper presented at a meeting at the Agricultural University of Norway, Aas, Norway, May 1973.
9. MP, Marius Pedersen Ltd., Unpublished data on vibration/sifting experiments in pilot scale; write to Marius Pedersen, Ferritslev, Funen, Denmark.
10. MUUS, E.B., Unpublished results of analyses of fodder pills from the Cecchini Rome Plant and from the pilot scale ESMIL experiments in Holland; write to Muus, Odense, Denmark.
11. TJELL, J.C., Chapter III, Vol. II, *Report on Land Application of Sludge* (in Danish with English summaries), in print, for publication 1981 by Polyteknisk Forlag, Lyngby, Denmark.
12. JENSEN, K.O. and SEVEL, T., *Mercury in Sludge in Relation to Danish Agriculture* (in Danish with an English summary), *Report on land application in sludge*, in print, for publication 1981 by Polyteknisk Forlag, Lyngby, Denmark.
13. MIN, *Minerals in Fodder* (in Danish) Landbrugets Informationskontor, Copenhagen, 1974.
14. KVÄRNER, Data on Cecchini fodder pill analyses, personal communication through Bjerre-Jorgensen, Marius Pedersen Ltd., Ferritslev, Denmark.
15. TJELL, J.C., House dust examinations, analyses for heavy metals, work in progress, Department of Sanitary Engineering, Technical University of Denmark, DK-2800 Lyngby.

Dr. Tjell holds an M.Sc. in chemical engineering and a Ph.D. on ion uptake in plant roots. After 2 years in Uganda on a Tse-tse fly control project, he returned to the Technical University of Denmark, where sludge utilization was the main research theme. He is now senior lecturer in sanitary engineering, specializing in solid waste.

Mr. Elmlund holds an M.Sc. in sanitary engineering. In 1978 he joined gendan inc. in Copenhagen, who have since 1974 encouraged recycling in Denmark and carried out research on food waste from private households.

Professor Jens Aage Hansen has produced some 60 publications in national and international journals, proceedings and conference reports, of which about 20 relate to solid waste and sewage sludge. He is a member of the Scientific and Technological Committee of the International Solid Waste Association, and has been on the EEC Advisory Committee on programme management, 'Recycling of Industrial and Household Waste' since 1979. He has acted as a consultant to the WHO and IAEA, and been visiting Associate Professor at the University of Washington Department of Civil Engineering, and is now Professor at Aalborg University in Denmark.

7 Recycling of Domestic Food Waste

Aage Heie and Johannes Minsaas

MOTIVATION

General concern regarding available resources emerge among the important preoccupations of humanity during recent decades. Consciousness of limitations in raw material supply and in the availability of various commodities leads to a growing understanding of the necessity for fair allocation of the world's resources. Increased efficiency in our use of raw materials has been brought into the limelight and the utilization of food waste for animal feed in this respect reveals an interesting challenge.

Interest in utilizing food waste for animal feeding generally increases during wartime. Under normal conditions, where concentrates are readily available, interest in utilizing ordinary household waste faded away. Only food waste from bigger institutions appears to be considered a resource of general interest.

A new approach originated in Italy some years ago, where Cecchini considered food waste contained in municipal refuse as a raw material rather than just offal. It is worthwhile noting that in Italy substantial amounts of dried garbage-derived animal feed is regularly marketed.

Norway has a deficit of grain which could hardly be covered by food waste. To feed the 4×10^6 population and the animals necessary to provide products traditionally considered necessary, 5-700 000 tonnes of grain is imported annually. As far as food waste can be recycled, therefore, this means providing work domestically instead of farm work in those countries growing the grain we otherwise needed to import.

The price system in Norway, according to the agreement between the farmers' union and the Government, involves a high levy on concentrates. High concentrate prices, in turn, mean high value on the farm for home-growing animal feeds. Since levies are unlikely to be asked for garbage-derived feeds, the price system therefore also favours recycling of food waste.

FOOD WASTE CHARACTERISTICS

Definition

The concept of 'food waste' may include all waste of animal and vegetable origin from a number of sources, both in production, refining and consumption of food.

The production waste is primarily farm waste, i.e. manure, straw and silage wastewater. Although these waste types usually contain considerable amounts of nutrients, they will not be included in the food waste concept in this connection.

In the treatment, packaging, canning and marketing areas of business, several by-products also arise. The amounts may be quite great, e.g. trimmings from the fish industry, trimmings, bones and bowels from slaughterhouses, whey from dairies and potato waste from potato flour factories. These by-products are ordinarily easily collectable in quantities that make re-use or recycling economic. The same is to some extent true for the marketing business and in large kitchens.

What we traditionally mean by 'food waste' is the waste that arises in the preparation and consumption of food in households. This waste includes both inedible parts of the food, uneaten food that the household does not care to save for another meal, and food that has deteriorated. This is the resource to be dealt with in this chapter.

Amounts of Food Waste

The amount of food waste from private households may be estimated by investigations on municipal refuse. This will not give the total amount, because some food waste (e.g. soups and sauces) is flushed into the sewage system, some is utilized as pet food and some is composted in gardens together with garden waste.

Nevertheless, the amount of food waste in the refuse will give a basis for estimating the resource potential of food waste.

SINTEF has investigated the amount and composition of domestic waste in several Norwegian communities. The results are given in Table 7.1. As can be seen, over one third of this is food waste. This includes also a minor amount of flowers and potted plants, which may also be considered as food waste in connection with recycling as animal fodder. In Norway, this food waste corresponds to 280 000 tonnes per year. For comparison, it may be mentioned that Norway's 1978 crop of turnip, swedes and kale grown for fodder amounted to 296 000 tonnes.

The food waste consists of different parts of the food and we have tried to

TABLE 7.1 Standard Norwegian Household Waste Amounts and Composition[1]

Component	Composition (weight %)	Amounts g/cap./day	tonnes/year*
Paper	33.5	168	245 000
Food	38.5	193	280 000
Textiles	3.1	16	23 000
Rubber and leather	1.1	6	8 000
Plastics	5.1	26	37 000
Other combustibles	5.3	27	39 000
Metals	4.0	20	29 000
Glass	5.1	26	37 000
Other non-combustibles	1.4	7	10 000
Fines (<10 mm)	2.9	15	21 000
Total	100.0	500	730 000

*Based on 4×10^6 inhabitants in Norway.

segregate it into various classes. The results are given in Table 7.2. About 45% arises as waste because of bad husbandry.

The composition of nutrients is given in Table 7.3. Here some results on food waste from big kitchens are also included.

TABLE 7.2 Food Waste Classes[1]

Food waste class	Weight % of food waste
Clean food waste due to wastage (meal leftovers, soups, sauces, meat, cheese, etc.)	30
Bread, cakes, uneaten sandwiches, etc.	10
Peel and leaves from fruit and vegetables	50
Others (egg shells, inedible trimmings and bones from meat and fish, etc.)	10

TABLE 7.3 Chemical Composition of Food Waste

Source	Dry matter (weight %)	Composition, weight %				
		Protein	Fat	Fibre	Carbonhydrate	Ashes
Private households						
fresh[1]*	32	16	11	4	57	12
sterilized[2]*	17	20	20	5	44	11
fresh[3]†	39	15	18	3	57	7
Big kitchens, sterilized[4]*	24	29	22		35	14

*Norway
†USA

Feed Potential

It is obvious that the great amount of food waste in the domestic waste represents a resource. Feeding experiments to be described later indicate that 2.5-3.0 kg of fresh food waste is equivalent to 1 feed unit, which corresponds to 1 kg of barley of good quality. If we assume 3 kg per feed unit, the amount of private household food waste in Norway represents about 90 million fodder units.

It is, however, unrealistic to imagine that all this waste would be available if a recycling scheme were introduced in Norway. As regards source separation, we shall see that some people are not willing to separate the food waste and the segregation efficiency is far below 100%. Further, collection and transport will be too costly to introduce source segration in large parts of Norway. As regards mechanical separation, the available plants require too large a population centre to cover more than a part of Norway.

But let us assume that half the amount is available for recycling. This corresponds to about 45 million feed units. If this were utilized as swine fodder, it would, theoretically, produce about 180 000 pigs or 13 500 tonnes of pork per year. The total consumption of pork and bacon in Norway is about 83 000 tonnes per year and it should thus be possible theoretically to save 16% of the feed to swine in the country.

The total consumption of feed in the Norwegian domestic animal production is about 3370×10^6 feed units. The food waste-derived feed units would make up only about 1.3% of this amount.

Another way of looking at this is that the feed unit amount corresponds to roughly 40 000 acres of agricultural land (as barley), which could be utilized for other agricultural production.

SEGREGATION OF FOOD WASTE

To utilize the great fodder potential of domestic waste, it is necessary to segregate, collect and bring it to treatment plants. The segregation can be done, in principle, in two ways, viz. at source in the household before it is mixed with the rest of the refuse, or in a central plant from mixed refuse.

Several systems for collecting and transport are possible. The principles of some are shown in Fig. 7.1. For central sorting we have the conventional system, but source segregation requires an extra garbage bin or central containers; or one can use, for instance, red bags for the food, as in the last system, which however did not work too well in an experiment employing a compactor lorry. The transportation may be done by two lorries driving the same route, separate container transport or by means of a special vehicle with two chambers.

All systems have advantages and disadvantages, which have to be evaluated both economically and by non-monetary parameters such as flexibility, working environment and so on. In a case study of Notodden, a small town with 13 000 inhabitants, it was found that the fourth system was the best. This was partly based on the fact that the feed processing plant was to be sited near the existing waste incinerator to make use of the recovered energy.

Efficiency of Source Segregation

The efficiency of source segregation is of course highly dependent on the population's ability to segregate the food waste. This type of segregation has a long tradition in Norway, but today is practised in only one town. This is partly due to the legislation and rules prescribed in 1964, which require sterilization. Most smaller waste utilizers could not afford this advanced technology and the system collapsed. However, it was partly due also to high manpower costs in Norway, which results especially in high collection and transportation costs. Today some 20 sterilization plants are in operation in Norway, and all but one collect their raw material from kitchens in hotels, restaurants, hospitals, and so on.

To investigate the attitude towards source segregation and its efficiency, a test program was run in Notodden. It lasted for 6 weeks during spring 1977 and 8 weeks during autumn 1977. A total of 302 households were asked to participate. Of these, only 17 refused categorically, while 14 were unwilling because of fear of extra work, nuisance of animal pests and so on. 27 already utilized the food waste as pet food or for composting, while 37 did not eat at home or were convinced that they did not produce any food waste. With some giving more than one reason for not participating, 224 households segregated their waste, corresponding to a participation rate of 74%.

The households were offered a collection device for the kitchen, either a plastic bag holder or a small plastic container, to be fitted inside a cupboard.

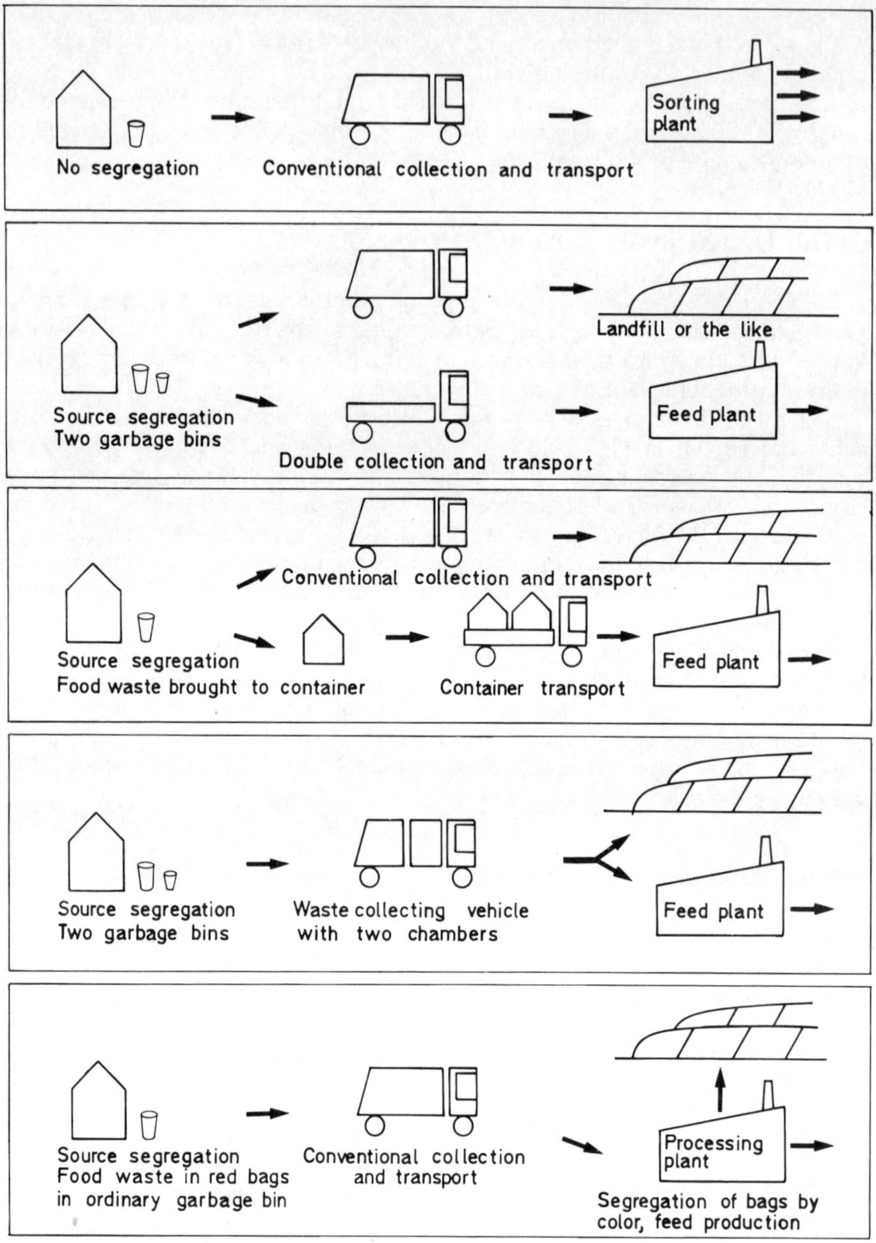

Fig. 7.1 Different systems for food waste segregation, collection and transport.

Also, they received a plastic 27 l bin to place beside the bin for ordinary refuse outdoors. The bins were equipped with plastic sacks and both bins and the sacks were marked a number to identify the household. Each week the food waste was collected and each sack weighed and related to the number of household members.

Before starting the test, we had investigated the waste composition in Notodden and we also investigated the waste in the ordinary garbage bins during the test periods to estimate the segregation efficiency. The results are summarized in Table 7.4.

The total amount of food waste was 195, 174 and 211 g/person/day, compared with 193 g in the standard waste (Table 7.1). About 65% of this was segregated. Taking account of the participation percentage, we obtained about 48% of the total food waste in the area under investigation.

TABLE 7.4 Segregated Food Waste and Composition of the Ordinary Waste Before and During Test Programme in Notodden

Waste component	March 1977		May-June 1977		Aug.-Oct. 1977	
	wt%	g/cap./day	wt%	g/cap./day	wt%	g/cap./day
Segregated food waste						
Ordinary waste	—	—	—	114	—	135
Food waste	46	195	21	60	29	76
Paper	26	109	41	119	26	68
Other components	28	119	38	109	45	120
Total	100	423	100	402	100	399
Segregation efficiency (%)	pre-test		66		64	

In Moss, where source segregation in private households is still practiced, we found that the yield was about 120 g/person/day, which agrees well with the Notodden results.

Mechanical Separation

The other alternative for domestic waste is mechanical separation. There are a number of separation plants in full operation throughout the world, but none in Norway. Further, we know of only two capable of isolating a fraction for animal feed; the Italian Cecchini system and the Spanish Enadimsa system.

The Cecchini process is based primarily on the particle size of the different waste components. As a consequence, the food fraction is polluted with other particles of the same size and has to be cleaned by a washing procedure. By sterilizing, drying and pelletizing, a feed is produced that is suitable for several domestic animals. It is calculated that such a plant will recycle about one third of the food waste, but the Norwegian licencee is at present working on adjustments of the process to fit Norwegian waste and, thus, to achieve a higher yield.

The Enadimsa process is based on the US Bureau of Mines process. Here, the waste is shredded and segregated by air classifiers and trommels. So far, neither the Bureau of Mines plant (US) nor the Enadimsa plant in Tournan en Brie (France) is actually producing an animal feed fraction, but quite optimistic results are claimed for the Enadimsa pilot plant in Spain.

QUALITY OF WASTE-DERIVED FEEDS

Feeding value and qualities

The feeding value of food waste reflects the chemical analysis and the digestibility of nutrients. Digestibility expresses the percentage of nutrients eaten that is utilized by the animal, and is determined through feeding experiments with species in question. Based on the digestibility of main nutrients and the chemical analysis, the energy value may be calculated to appraise the economic value.

Digestibility of food waste for pigs has been determined in Norwegian experiments. Some results are shown in Table 7.5, where the figures for barley are also shown for comparison. In the wartime food waste, the protein and fibre digestibility, in particular, was inferior compared with today's food waste. It also seems that drying lowers the digestibility, but these were experiments that were not 100% successful, so it may be possible to raise these figures.

TABLE 7.5 Digestibility of Food Waste for Pigs

	Coefficient of digestibility					
	Dry matter	Organic matter	Crude protein	Fat	Carbro-hydrate	Fibre
Boiled, 1945[5]	81	82	71	89	87	58
Sterilized, 1978[2]	86	89	86	86	94	77
Sterilized & dried 1978[2]	82	83	65	82	92	47
Barley[6]			75	76	92	41

Based on digestibility data in Table 7.5, it is calculated that 2.5-3.0 kg of fresh mixed food waste represent 1 feeding unit (the feeding value of 1 kg of standard barley, 1 650 net kcal or 6930 kJ).

Analytical data for food waste are scarce in the literature, particularly regarding elements of nutrition. The content of crude protein, lipids and carbohydrates is fairly well established, while little is known regarding amino acid and lipid elements. The crude protein content of food waste is high enough to be adjusted by introducing feed grain with low protein when developing a ration for economical feeding. Such practice probably demands adjustments also of the amino acids of the ration to secure favourable feed conversion.

Feeding food waste in combination with low-fat feed grains seems favorable for the adjustment of fat in the ration. Food waste is high in fat and since the fat is of a quality that varies and is generally unfavourable for pork quality, food waste should not be fed liberally. By adapting feeding technique and feeding restricted amounts of food waste, however, quality pork can be produced.

The pork quality from pigs fed food waste has been investigated in a joint enterprise of the Agricultural University of Norway and SINTEF. The results are summarized in Table 7.6. It appears that, when using relatively moderate amounts of food waste, carcasses with satisfactory fresh quality are produced. Organoleptic criteria are favourable. Chemical criteria, however, indicate somewhat doubtful fat quality for cured products and for extended deep freeze storage. For these purposes TBA figures of fat, for instance, should not exceed 220.

The results reveal that what was considered a moderate food waste level appears to be too high when raising pigs for curing. Probably the feed energy from food waste should not exceed 20-40% of the total feed energy.

Both poultry and cattle products are less sensitive to feed fat quality compared with the pig. In the USA[3], the use of dried food waste is reported for these species. Another possibility is the production of a defatted dried food waste product and disposal of the fat for purposes similar to those where crude meat and bone fat is used, for instance for poultry feeds.

TABLE 7.6 Quality of Pork from Pigs Fed Different Amounts of Sterilized Food Waste[2]

	Control	Food waste feeding	
		Moderate	High
% feed energy from food waste	0	40	70
Color[a]	2.3	2.2	2.2
Firmness[b]	11.7	10.3	10.6
TBA[c]	210	360	380
AOM[d]	2.7	2.3	1.7
Iodine value[e]	63	69	70
Organoleptic quality[f]	8.9	8.8	8.7

[a]Meat color: 0-5 points, 0 light, 5 dark color, 2.5 desirable.
[b]Back fat firmness: 0-15 points, 9 minimum acceptable level.
[c]Thiobarbituric acid number: expresses the content of aldehydes in quite fresh fat. High count indicates low stability and risk of rancidification.
[d]Active oxygen method: expresses time under fixed conditions to develop a certain degree of rancidity. High count means stable fat.
[e]Iodine value expresses the degree of unsaturation of the fatty acids.
[f]Flesh taste: 0-10 points, 6 minimum acceptable level.

Dried full-fat food waste could be used without problems in the compounding of poultry feeds and probably also in compounding cattle feeds. Processed food waste, whether wet or dry, seems to be found agreeable by all kinds of livestock and pets.

Heavy Metals and Other Unwanted Substances in Food Waste

Food waste manually segregated from municipal waste in Trondheim did not contain serious amounts of heavy metals[1], although it contained more than reported for source-segregated food waste in Chapter 6. The analytical data for heavy metals in food waste are scarce. However, work is being performed in Denmark and Norway on both source and mechanically segregated food waste. In judging the results, it must not be overlooked that many elements are indispensable and often added to commercial feeds to adjust them to recommended levels.

Moulds grow on moist organic material and growth conditions on food waste ought to be favourable. Such moulds in animal feeds might do harm at least three different ways:

More or less poisonous substance (mycotoxins), which might cause disease, are produced.
Through growing on special components, moulds may cause an imbalance among available nutrients.
Moulds may be a direct cause of pathological conditions, so-called mycoses.

At present, there is hardly reason to believe that moulds growing on food waste produce hazardous amounts of mycotoxins. Fermentation in the waste produces organic acids which depress the growth of moulds. If food waste is going to be widely adapted in animal feeding, better knowledge of mould growth and mycotoxins will be demanded.

PROCESSING OF FOOD WASTE

Direct feeding of food waste to pigs, for instance, would be the simplest possible way of converting waste to feed. This practice has been illegal in Norway since 1964 and is not too encouraging with respect to optimum utilization of feed waste nutrients.

Industrial processing demands collection of food waste for bigger treatment units. The most advanced processing, apparently, would be drying and pelletizing. SINTEF has a project in progress in this field but it is still too early to report the study. It is anticipated that dried food waste will be an excellent straight for compounding.

Regardless of using the food waste wet or dry, processing has to commence by approved sterilization, which has since 1964 meant batch treatment for a minimum of 20 min at a temperature of 120 °C.

Up to now, any continuous process has been rejected. However, such processes are steadily improved and, therefore, sooner or later approval must be seriously discussed with authorities.

Sterilized food waste may be used for feeding in liquid or dried form. Since a sterilizing unit could operate economically with quantities too small to occupy a dryer of optimum size, food waste may be collected from several local sterilizing units and brought to a central dryer.

To ensure the best possible utilization of nutrients contained in the food waste, each step during processing should be gentle.

Possible Ways of Using Liquid Food Waste

Following disintegration and sterilization, food waste forms a heavy slurry with quite a pleasant odour. The slurry is obviously palatable to livestock. In greater quantities, it would be inconvenient to feed the product this way and in modern poultry husbandary quite impossible since dry feed automats are adapted.

Herdsmen[7,8] are stressing the problem of mastering the high fat content of food waste in practical feeding conditions. The high dry matter fat content and the changing quality of fat, which is beyond the herdsman's control, is considered the negative side of food waste as pig feed.

Distribution of liquid food waste would call for acceptable preserving agents if not used directly under the sterilizing plant. Possible preserving agents could be formic acid or sodium meta-bisulphite ($NaHSO_3$). Biological acid preservation by means of acid fermentation might also be applied, possibly aided by the addition of, for example, molasses.

Liquid food waste could form a useful 'carrier' for liquid feed supplements, possibly in combination with molasses. Liquid feeds are rapidly increasing in cattle feeding.

Another possible use of liquid food waste appears to be as a pellet binding agent. Preliminary tests conducted by SINTEF (not yet published) produced pellets of outstanding physical quality when small amounts of food waste were used. In these tests, pellets were produced without steaming the flour prior to pelleting.

Dried Food Waste

By development of an adapted drying technique and possible modification of suitable dryers, the careful drying of food waste should be possible without serious depression of digestibility figures. The Ministry of Environmental Protection finances a drying project which is supervized by SINTEF and carried out in collaboration with producers of drying equipment.

The project is still in progress, but experiences so far are encouraging. It is expected that before the end of 1980 SINTEF will be able to advise regarding dryers, their modification and methods for drying food waste.

Drying is considered the safest way of converting food waste to a commodity for the feed trade, partly because drying and pelletizing solve stocking problems but mainly because a dried product easily 'vanishes' when properly used in small amounts in the compounding industry.

Aiming at the animal feed industry and the quality traded, the processing and inclusion of available amounts of food waste should easily be adapted without hazard to the general quality of animal products. On a dry matter base, the practically available food waste in Norway would hardly exceed 45 000 tonnes annually or approximately 2.5% of the quantity of concentrates used during 1978.

ECONOMY FOR FOOD WASTE RECYCLING

Source segregation, separate transport and processing of food waste is quite expensive. In Notodden, the case study city of SINTEF's project, the population of 13 000, big kitchens and some minor food processing factories are producing some 800 tonnes of food waste per year or about 300 000 feed units. The investment in a processing plant including dryer is calculated to 1.2 million NKr (1978) or about $240.000. The collecting and transport is supposed to be done by a combined lorry and the steam required is supposed to be delivered free from the existing waste incinerator. The total cost per feeding unit is estimated to 2.20 NKr, or about $0.44. The per cent contribution of the different factors is given in Table 7.7.

The feeding unit price for concentrates is in fact much lower. The price system in Norway, however, involves a high levy on feed concentrates, so that the current price is about $0.45.

TABLE 7.7 Contribution of the Different Cost Factors for a Food Waste Recycling Scheme in Notodden

Factor	% contribution
Capital costs, collecting bins and food processing plant	28
Electricity and water	3
Maintenance	9
Wages, processing plant	14
Collecting and transport of industrial food waste	7
Extra expenses for the municipal collecting and transport (capital costs, wages and food waste sacks)	39

REFERENCES

1. NATVIG, K., KNAP, A.Hj. and HALMO, T., *Municipal Waste III*, STF21 A76029, NTNF-UFA No. 1.1.14, SINTEF Trondheim (1976) (in Norwegian).
2. LYSØ, A., *Food Waste as Feed for Swine*, Preliminary report to the

Commission for Animal Feeding Trials, NTNF-UFA No. 2.1.23, NLH, Aas (1978) (in Norwegian).
3. MORRISON, F.B., *Feeds and Feeding, 21st ed.* Morrison Publishing Company, Ithaca, New York, USA (1951).
4. NUNN, D.W. and STROM-ERICHSEN, A., *Use of Waste from Food Production and Consumption,* CMI-74176-1, NTNF-UFA No. 1.3.11, CMI, Bergen (1976) (in Norwegian).
5. HUSBY, M., *Reports from the Agricultural University of Norway* (NLH),*25* (1945) (in Norwegian).
6. BREIREM, K. and HOMB, T., *Animal Feeds and their Preservation,* Buskap og Avdraatt A/S Gjøvik, Norway (1970) (in Norwegian).
7. JOHNSRUD, M., *Practical Experiences with sterilized Food Waste in Pig Feeding,* paper read at the 'Concentrate Day', Pfizer A/S, Oslo (1966) (in Norwegian).
8. HOMB, T., *Feeding Value of Kitchen Waste: Use of Kitchen Waste in Swine Feeding,* paper read at the 'Concentrate Day', Pfizer A/S, Oslo (1966) (in Norwegian).

Dr. ing. Aage Heie is presently, after some years with chemical and biochemical research, leader of the 'Garbologists' (Garbage Scientist Group) at SINTEF in Trondheim, Norway. SINTEF is a non-proft organization performing research under contract to corporations, industrial association, public service agencies, government departments and other clients, and has a staff of about 950, of which about 10 can be classified as garbologists. Dr. Heie's main projects at SINTEF have dealt with recycling of food waste, treatment and recycling of household, industrial and hazardous waste, regional solid waste management in rural areas, master plan for recycling in the Nordic countries and solid waste management in the Suez Canal Region, Egypt.

Dr. J. Minsaas is Special Project Manager at the Foundation of Scientific and Industrial Research (SINTEF) at the Norwegian Institute of Technology, Applied Chemistry Division. He has held various posts in the compounding industry, and has been Research Director in animal nutrition at the Norwegian Herring Oil and Meal Industry Research Institute. His main interest is the reduction of wastes by turning them into useful by-products for recycling.

8 An Energy Balance for Plastics Recycling

J. F. Ingen Housz

INTRODUCTION

The question has been raised whether the plastics in municipal solid waste (MSW) should be recovered in order to be re-used as a material from which useful products may be made, or whether those plastics should be converted to a fuel.

In the case where plastics, together with (a part of) the paper and board fraction and other combustibles, is to be used as a fuel, either in bulk or as a densified refuse-derived fuel (dRDF), separation need not go as far as in the case where plastics and paper are to be re-used as materials.

The economic value of recovered plastic material depends on the quality of the products that may be made from such a material. For wood-replacement applications, separation of the various constituent materials — mainly polyolefins, polystyrene and polyvinylchloride — and cleaning need not go very far. The value of such material is low, however, as wood is a relatively cheap construction material. For more sophisticated applications, separation and cleaning must go much further, making the process more expensive and energy consuming, in order to arrive at a material that meets agreed specifications.

Costs as well as revenues for the separation and cleaning operations will therefore depend on the specification of the recovered material. An economic analysis of the various possibilities will be necessary before a decision is taken, and energy considerations play a very important role in this analysis. As the useful life of a plant is much longer than the period over which a reasonable prediction of energy costs can be made; it appears useful to make an energy balance for the various possibilities in addition to the economic analysis.

ENERGY CONTENT OF PLASTICS MATERIALS

In contrast to most other materials, the production of plastics materials requires energy not only in the form of process energy but also as the feedstock to the process. Nearly all plastics are oil derivatives, and oil represents energy.

Plastics in waste contain energy recoverable by combustion to roughly the same extent that oil has provided the feedstock. In the case of combustion the process energy is not recovered and must therefore be considered the price paid for the useful services which the products have rendered before being turned to waste. In the case of recovery of plastics as a construction material the process energy is also recovered, against which the extra energy consumption for the recovery and conversion processes must be considered.

For the three most common plastics an average energy content has been calculated from available process and statistical data by the Plastics and Rubber Institute TNO at Delft in 1976[1]. The results account for feedstock losses in the processes and for all energy inputs from extraction and transport of the crude (basis Persian Gulf) through polymerization. Raw materials and energy of plant, buildings, etc., are not included.

TABLE 8.1 TNO Estimates of Energy Content (TOE/tonne)[1]

Electricity used is converted at 4000 kWh/TOE (tonne oil equivalent). At an average of 42 GJ/TOE this means that a conversion efficiency of 34.2% has been used.

Chapman[2] arrives statistically at a figure of around 26% for the UK. US sources often quote 33%. Process steam is put at 12.5 tonne steam/TOE. For steam of average 2720 kJ/kg = 650 kcal/kg, this amounts to an 81% conversion efficiency.

These figures appear reasonable for our purposes. Actually, higher conversion efficiencies are possible, e.g. by using counter pressure turbines, in the case of combined steam and power generation, as is the usual practise at refineries.

	LDPE	PS	PVC
Feedstock	1.08	1.12	0.53
Process energy	0.93	1.10	1.20
Total	2.01	2.22	1.73

When plastics are used as fuel, the heat of combustion[3] is recovered. The remainder of the total energy content, as given in Table 8.1, is not recoverable by combustion. This non-recoverable part is shown in Table 8.2.

TABLE 8.2 Recoverable Energy of Plastics (TOE/tonne)

	LDPE	PS	PVC
Total energy content	2.01	2.22	1.73
Heat of combustion	1.10	0.99	0.48
Non-recoverable by combustion	0.91	1.23	1.25

Separating a plastics fraction from MSW makes re-use from the energetic point of view profitable if the 'use value' of this fraction is higher than its heat of combustion; in other and rather contradictory words, if a part of the non-recoverable energy content is recovered.

WASTE SEPARATION

In dry separation processes it is usual, after preliminary operations, to use air classification as the means to divide the waste into a light and heavy fraction. In case the light fraction is subsequently processed to RDF, the adjustment of this air classification may differ from the conditions used for the case of further separation to a reclaimed plastics fraction.

Both cases have the energy consumed in the front end of the process in common, and for a comparison of these two possibilities this energy does not enter into the analysis. Of course, if the choice were between mass incineration and RDF production, this energy should be considered.

For the purpose of this paper we use the flow sheet for a 2 × 20 tonne/h separation plant under construction in Vienna, Austria and designed by Esmil International[4]. The Esmil system (Fig. 8.1) yields a plastics fraction consisting almost entirely of polyolefin film, containing only small amounts of polystyrene, paper, straw, etc. PVC is nearly quantitatively removed at the heavy/light air classification. In this plant, the light fraction is separated into a compostable fraction, paper and plastics. The energy consumption for this separation has been attributed to each of the three fractions on a dry weight basis by considering each step in this further separation process separately and dividing its estimated energy demand over the fraction issuing from such step. This has included some guesswork as to the moisture contents of the various fractions in the process.

As a result, for 540 kg/h of plastics waste (from one 20 tonne/h line) the power requirement is 46kW and the use of process steam amounts to 340 kg/h. These figures convert to a separation energy input of 0.073 TOE/tonne plastics.

Of course the attribution on a dry weight basis to the three products is highly arbitrary. If the total energy demand of each of the steps which are necessary for the production of the plastics fraction were assigned to the plastics, then 140 kW and 2000 kg/h steam for 540 kg/h of plastics result in 0.36 TOE/tonne. This method applies for the case in which the compostable fraction and the paper fraction have no commercial value other than its value as RDF.

CONVERSION TO A PRODUCT

The specification to which a plastics waste fraction must be produced will depend on the product to be made from such a material. A straight comparison is best made for a product normally produced from virgin plastics material. The household refuse bag, used in many European countries, is a good example of a product suitable for such a comparison as the main quality aspect is mechanical strength, and properties as appearance, printability, colour, etc., are relatively unimportant. Lower mechanical strength of a waste-derived material may be compensated by increasing film thickness. The film blowing process, however, requires a rather good material homogeneity. Such refuse bags are produced normally from a low-density polyethylene of melt index 2, at a nominal wall thickness of 0.055 mm. Bags from high-density polyethylene exist but will not be considered here. Industrial scrap from heavy sack production, melt index about 0.3, is often used to replace a part of the virgin material. This requires a regranulation of the scrap. Waste film, separated at source, e.g. used shrink film and used heavy sacks, is also used; this material requires more extensive preparation before it can be re-used. These reclaimed materials are usually blended with virgin material in order to increase the melt index of the mix. The lower average melt index results in a stronger film than that made from virgin material only.

It is claimed that an Italian firm produces household refuse bags from a plastics fraction reclaimed from MSW. We have not been able to confirm this rumour and details about the product are consequently not available. From inspection of film blown from the plastics fraction produced at Esmil's pilot plant, we estimate

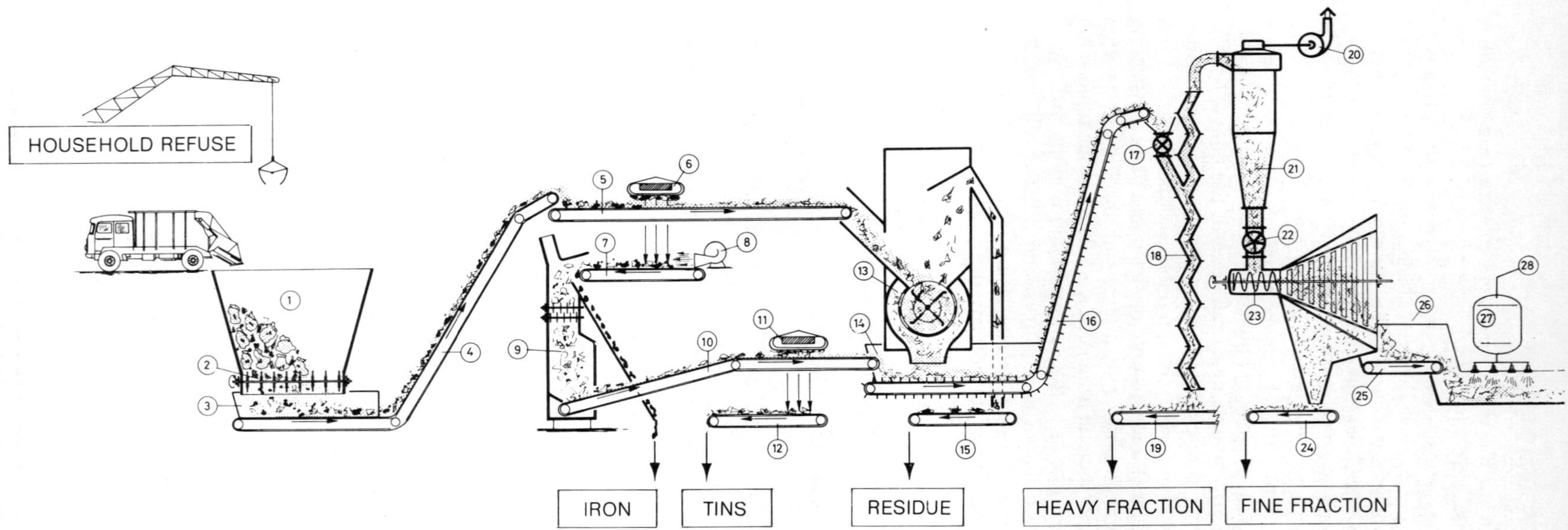

1 Feeding bunker
2 Shredder for plastic bags
3 Feeding bunker
4 Belt conveyer - 1
5 Belt conveyer - 2
6 Magnetic separator
7 Belt conveyer - 3
8 Air separator
9 Shredder for tins
10 Belt conveyer - 4
11 Magnetic separator
12 Belt conveyer - 5
13 Hammer mill (crusher)
14 Feeding bunker
15 Belt conveyer - 6
16 Scraper conveyer
17 Rotor lock - 1
18 Zig-zag classifier - 1
19 Belt-conveyer - 7
20 Fan - 1
21 Cyclone - 1
22 Rotor Lock - 2
23 Rotating screen
24 Belt conveyer - 8
25 Belt conveyer - 9
26 Conveyer chute
27 –
28 –

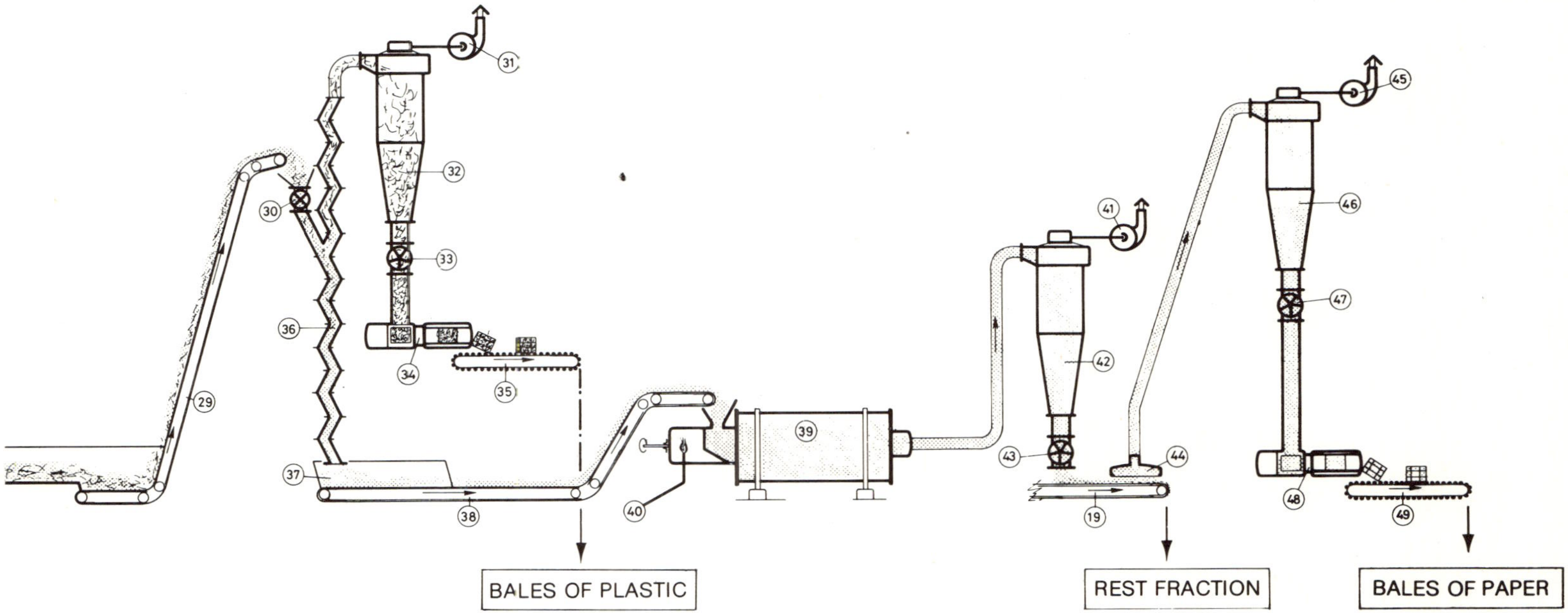

29 Belt conveyer - 10
30 Rotor Lock - 3
31 Fan - 2
32 Cyclone - 2
33 Rotor Lock - 4
34 Baler for plastic
35 Roller conveyer
36 Zig-zag classifier - 2
37 Feeding bunker
38 Belt conveyer - 1
39 Dryer

40 Gas burner
41 Fan - 3
42 Cyclone - 3
43 Rotor Lock - 5
44 Hood
45 Fan - 4
46 Cyclone - 4
47 Rotor Lock - 6
48 Baler for paper
49 Roller conveyer - 2

Fig. 8.1 The Esmil Household refuse separation system. (Reproduced by permission of Esmil International BV).

that for household refuse bags from such material a film thickness of nominal 0.075 mm will be required.

As all products made, from whatever input material, still contain the recoverable heat of combustion which will become available as and when such material is eventually burned, this recoverable energy does not as a rule enter the analysis. Only where material losses occur in such a way that the material cannot be directed to an RDF stream should we account for this loss. This is the case, for example, for material contained in screen packs that have been changed. Floor sweepings, material collected in dust separation cyclones, etc., which could be assumed to be available for burning, will also be considered completely lost. In normal manufacturing processes, this loss may be neglected. In the processing of scrap and waste, however, such losses may run from about ½% to over 2% and in such cases the recoverable energy lost must be included in the analysis.

PRODUCTION OF HOUSEHOLD REFUSE BAGS

Virgin Material

The conversion of virgin LDPE to household refuse bags involves the steps of extruding, blowing and cooling the film. Subsequently the lay flat film is sealed, cut and rolled into bundles of 20 bags. The complete process requires about 0.6 KWh/kg of input material and the overall material efficiency if 94%. The 60 kg/tonne input of waste material is recovered from the process at its heat of combustion value. The non-recoverable energy content of the final product is thus calculated as follows.

	TOE/tonne
non-recoverable energy of input material	0.91
conversion at 600 kWh/tonne	0.15
total in product	1.06
at 94% yield, 1.06/0.94	1.13

Note: as the 60 kg/tonne production scrap will as a rule be reprocessed, it may be argued that it should be recovered from the process at a higher value than its recoverable energy equivalent. Crediting this material at 0.72 TOE/tonne will reduce the figure of 1.13 TOE/tonne derived above to 1.09 TOE/tonne.

Industrial Waste

Such waste results from off-gauge film, misprints, production overruns, faulty seals and other sack-making errors, etc. This material is usually obtained through established trade channels. Transportation is involved, which we estimate at 0.036 l fuel/tonne km. At an average of 150 km, this amounts to 0.055 TOE/tonne if 1 m^3 of Diesel fuel is assumed to contain 1 TOE.

The treatment of this waste consists of cutting, pneumatic transportation and re-extrusion to granules or chips, using fine mesh screens to trap sand and dust. At 500 kWh/tonne input and an assumed overall material loss of ½%, the total energy cost of the reclamation process is:

	TOE/tonne
for transportation	0.055
for treatment	0.125

for materials loss	0.0055
total per tonne input	0.1855

At the overall yield of 99½%, energy input is 0.186 TOE/tonne. In the subsequent bag production, this material replaces its weight of virgin material as the film thickness remains the same. Conversion requires no extra effort as the slightly lower melt index has no noticable effect on energy consumption and production efficiency is not affected. The reclaim material therefore represents, at input to the process, the same energy as virgin material. The scrap, ex works, therefore represents 0.91 − 0.186 or 0.72 TOE/tonne.

Film Waste Separated at Source

This is material collected by the trade from users; main constituents are shrink film, heavy sacks and other packaging film. As a rule agricultural film is not included. Melt index is on average higher than for industrial waste. Treatment is the same as for clean industrial waste with an additional washing and drying operation. At a 100 tonne/week reclamation plant, where the air is heated electrically, the additional energy for washing and drying is 0.4 kWh/kg input. Total material loss amounts to 1%.

Total energy cost for reclaiming this waste are:

	TOE/tonne
transportation	0.055
cutting and re-extrusion	0.125
washing and drying	0.100
materials losses	0.011
total per tonne input	0.291

At 99% yield this amounts to 0.294 TOE/tonne reclaimed material. Again, this material replaces virgin resin on a weight for weight basis. Its energy value at input to the film blowing process is therefore equal to that of virgin material or 0.91 TOE/tonne, resulting in 0.62 TOE/tonne for the waste at source.

Note: because of its higher melt index, the reclaimed source-separated material requires less blending of virgin material than the reclaimed industrial scrap. The use of this material may therefore result in a lower cost price for the final product. Its economic value is therefore somewhat higher than that of the reclaimed industrial scrap.

Plastics from Municipal Waste

The material ex waste-separation plant must be given a treatment at least equivalent to that of waste separated at source. Thereafter it may (as assumed above) be processed to refuse bags at a film thickness of 0.075 mm.

The energy cost of the total operation, including the waste separation, is then:

	TOE/tonne
waste separation	0.073
transportation	0.055
cutting and re-extrusion	0.125
washing and drying	0.100
materials losses	0.011
total per tonne input	0.364

At 99% yield this results in 0.368 TOE/tonne for the reclaimed granules. Adding 0.22 TOE/tonne to this figure for conversion of the granules to bags, assuming the same overall efficiency as for virgin material, results in a total conversion cost, from waste to bags, of 0.59 TOE/tonne. One tonne of 0.075 mm film now replaces 0.733 tonne of 0.055 mm film from virgin material. Its energy value is therefore equal to only 0.733 × 1.13 = 0.829 TOE/tonne. The non-recoverable energy of the plastics in the original waste is therefore recovered to the extent of 0.83 − 0.59 = 0.24 TOE/tonne.

If the higher energy figure of 0.36 TOE/tonne for the separation is used, the result is less encouraging:

separation energy for plastics fraction	0.073	0.360
energy used for reclaimed granules	0.368	0.657
total for conversion to bags	0.59	0.88
equivalent value of bags	0.83	0.83
net gain or loss	+0.24	−0.05

This result shows that our analysis is rather sensitive to the method in which energy used in the separation plant is assigned to the various products recovered.

CONCLUSIONS

Based on the non-recoverable energy content of low density polyethene, an 'energy value' of various types of scrap and waste has been derived as follows:

	TOE/tonne
virgin material	0.91
industrial scrap	0.72
waste separated at source	0.62
waste contained in MSW	
high	0.24
low	−0.05
RDF	0.00

The heat of combustion of the materials is the same throughout; if RDF were to be assigned a lower 'recoverable energy' value than this heat of combustion, the difference would have to be added to the other materials as this would in effect increase the non-recoverable energy of the virgin material, which is the starting point for these calculations.

Such a lower value could be argued on the basis of the unavoidable and usually rather high moisture content of RDF[5]. Wilson[6] uses steam-raising efficiencies for RDF and for coal of 0.69 and 0.81 respectively, resulting in a substitution equivalent of 1.17. This means that 1 tonne of LDPE in RDF, having a heat of combustion of 1.10 TOE replaces 0.940 TOE in coal. Using this figure, the non-recoverable energy content of virgin LDPE would increase to 1.07 TOE/tonne and the energy value of LDPE film in MSW would then be 0.40 TOE/tonne.

The figures for energy content of plastics are based on pre-1976 data. More modern plants may well lower these figures, tipping the balance in favour of RDF. Against this is the fact that waste separation on an industrial scale is in its infancy and improved energy efficiency may be expected in future plants.

Even though 'lart de grouper les chiffres' is not foreign to this exercise, the conclusion appears justified that, if in any separation set-up the recovery of plastics is economically viable, future increases in the cost of energy will tend to strengthen this position.

A second and more obvious conclusion is that separation of plastics wastes at source recovers a much larger part of the process energy, used in making the plastics material, than the more devious recovery route via the separation of plastics from MSW.

Finally, if the paper fraction is used for RDF, the plastics had better be burned also, as the total separation energy in that case has to be assigned to the plastics fraction, resulting in an energetic loss instead of the intended gain.

REFERENCES

1. KRI-TNO, Report 323/'76 part VI-Procesdata.
2. CHAPMAN, P. F., *Energy Policy* **2** (1974) 91.
3. AGGARWAL, S. L., *et al.*, Chapter 5 in *Polymer Handbook, 2nd edn.*, BRANDRUP, J. and IMMERGUT, E. H. eds., Wiley, New York (1975).
4. Esmil International, Project 66.0000, Amsterdam (1979), private communication, Mr. Hoogendoorn.
5. SHENG, H. P. and ALTER, H., *Resource Recovery and Conservation* **1** (1975) 85.
6. WILSON, D. C., *Resource Recovery and Conservation* **4** (1979) 161.

Professor J. F. Ingen Housz graduated in Mechanical Engineering at the Technical University of Delft. After three years in the USA with Babcock & Wilcox and E.I. Due Pont de Nemours, he joined the Van Leer Group of Companies in 1954. In 1965 he was appointed to his present position in the Department of Mechanical Engineering at the Technical University of Twente, where he teaches Polymer Technology. His main research interest is in extrusion of thermoplastics. He is chairman of the advisory councils of the Plastics and Rubber Institute TNO and of the Packaging Institute TNO and chairman of the executive committee of the Foundation for Packaging and the Environment; it is in this latter capacity that he became interested in processing problems connected with the recovery of plastics.

9 Materials Recovery by Central Sorting of Household Refuse

A. G. Buekens

INTRODUCTION

Survey

Values contained in household refuse can be recycled either after collection of source segregated materials or after centralized manual or automated sorting of refuse.

This chapter reviews available equipment and systems for the central sorting of household refuse and briefly discusses the market potential of the obtained fractions. Similar methods have been tested to classify incinerator residue, shredded cars, appliances, machinery, old tyres or cable scrap.

Classification of Reclaimable Wastes

Wastes can be classified into production, distribution and (post-) consumer wastes. *Production wastes* are normally much better defined and more homogeneous than consumer wastes and hence are easier to recycle. They can be exemplified by the following:

1. *Home scrap* is generated during the casting of metal into ingots. It is readily identifiable, almost uncontaminated and can often be directly recycled into the production process. The same holds for broke in a paper mill, cullet in a glass factory or thermoplastic waste at a plastics processing plant.
2. *Prompt* or *new scrap* arises during machining, boring and cutting of metal. It is also identifiable and can be delivered to the reclamation industry in a segregated form, only slightly contaminated with oil and dust.
3. *Obsolete* or *old scrap* consists of structural steel, scrapped machinery, wrecked cars, etc. It can be recovered as long as market conditions and operating costs allow for profitable recovery. *Commercial* and *distribution* wastes consist mainly of packaging materials; they are recoverable where local segregation and storage are feasible. *Consumer* wastes are composed of household refuse, bulky wastes, cars, tyres etc.

Apart from an occasional boom, as during the Korean War or the years 1973/74, it has been difficult in industrialized countries to integrate household waste into an economically viable circuit: refuse is composed of an almost infinite variety of products, which have in common only that they are no longer wanted. The variable size and form of their components, the unpredictable erratic composition, the presence of all kinds of contaminants and of 'pernicious contraries' challenge any sorting method and form a source of uncertainty for any type of automated operation and for product quality.

Forms of Recycling

The recycling of wastes is one of the oldest human activities. It may assume different forms:

1. *re-use* of rejected articles, such as machinery, cars, appliances or clothing, often after previous examination, repair and cleaning,
2. *closed-loop recycling*, i.e. the re-use of waste materials in applications similar to the original ones, e.g. making new glass containers from cullet or new steel from scrap,
3. *open-loop recycling*, i.e. the re-use of waste materials in lower grade applications, e.g. making board from waste paper or reinforcing bars from mixed steel scrap,
4. *conversion of wastes* into new materials, such as fuel, compost, animal fodder or building materials.

The recycling potential of wastes normally depends on economic factors, such as available quantities and grades, market value of the reclaimed products, available reclaiming equipment and — not least — labour cost.

The Reclamation Industries

The reclamation industries have a remarkable and jealously guarded know-how in the tracking, purchasing and processing of recoverable wastes and in vending reclaimed products. However, they are much more vulnerable than other industries because:

1. wastes are available only in limited, variable and rather unpredictable quantities and qualities, so that continuous, large-scale activity based on long term contracts is difficult to achieve;
2. investment and maintenance cost is high per unit volume of treated material, because of the fluctuating supply of material and the relatively small scale of operations;
3. salvage and reclamation are difficult to automate and labour-intensive;
4. social acceptance of reclaimed products and of reclamation activities is low; plants and stockpiles will often cause aesthetic offence and air, water or soil pollution;
5. many specifications still demand the absence of any reclaimed materials in the product.

The reclamation industries are subject to more intense cyclic variations than any other branch of industry. During a boom, the value of reclaimed products increases rapidly, but so does competition. During a slump the value of stockpiles drops, disrupting the financial equilibrium of the reclamation enterprises, many of which go bankrupt.

Nevertheless, reclamation supplies large tonnages materials to the metal, the paper and board, the glass, and the plastics industries.

UNIT OPERATIONS IN DRY SORTING

Survey

The oldest, simplest and most widely used sorting method is the handpicking of valuable components from waste. The sorter may be a tramp, looking in the dustbin for edible or saleable material, men, women or children scavenging on a waste dump, or workpeople, sorting material from a conveyor belt in a refuse-disposal or a waste-sorting plant. Although most of these activities will meet criticism on hygenic, social and aesthetic grounds, it should also be stated that recovery on the dumpsite is an organized and basic part of developing economies all over the world.

The capacity of sorting at a belt varies from a few hundred kg/h for light and bulky material to several tonne/h for dense material. Typically the belt moves at a speed of 10 m/min. When market conditions are favourable, the sorters will recover enough material to pay for themselves.

Manual sorting is important also in the classification of miscellaneous materials and objects and in the removal of contraries from a saleable product. These contraries may be valuable, e.g. in the removal of aluminium caps or brass from lamp bulbs from cullet.

Mechanized and automated sorting systems have been developed in ore beneficiation, coal washing and fruit sorting. The sorting operation is based on difference of dimension, shape, density, elasticity, brittleness, friction, colour, magnetic susceptibility, reflection of light, radioactivity, etc.

Separators normally require feed material within a particular size range, obtained by a suitable combination of crushing, screening or even agglomeration.

Sorting methods can be subdivided into dry and wet methods. Dry methods are cheaper in investment and operating cost than wet methods, but they may create dust problems and explosion or fire hazards. Moreover, the recovered fractions are still contaminated with dirt and organic material. Wet methods are more expensive; they yield a cleaner product, but give rise to a waste water treatment problem. Table 9.1 lists a number of dry and wet methods, with their size range of application.

In refuse sorting it may be necessary to liberate the waste from plastic or paper bags, reduce the size of bulky objects, disentangle mixed and disintegrate heterogenous materials.

Yields and Grades

Each sorting operation subdivides a feed stream into at least two fractions, between which the components of the original mixture are distributed. The performance of a sorting operation on a component A can be characterized by a concentration ratio, a yield and a product grade.

When a magnetic separator, for example, is used to extract ferrous metal from a refuse layer, spread onto a conveyor belt, the previous factors can be defined as follows:

$$\text{Yield} = \frac{\text{Amount of ferrous metal in the concentrate}}{\text{Amount of ferrous metal in the feed stream}}$$

$$\text{Product grade} = \frac{\text{Amount of ferrous metal in the concentrate}}{\text{total amount of concentrate}}$$

$$\text{Concentration ratio} = \frac{\text{Flow rate of feed material}}{\text{Flow rate of concentrate}}$$

A high yield, product grade and concentration ratio are desirable, but also

TABLE 9.1 Solid-Solid Separation Methods

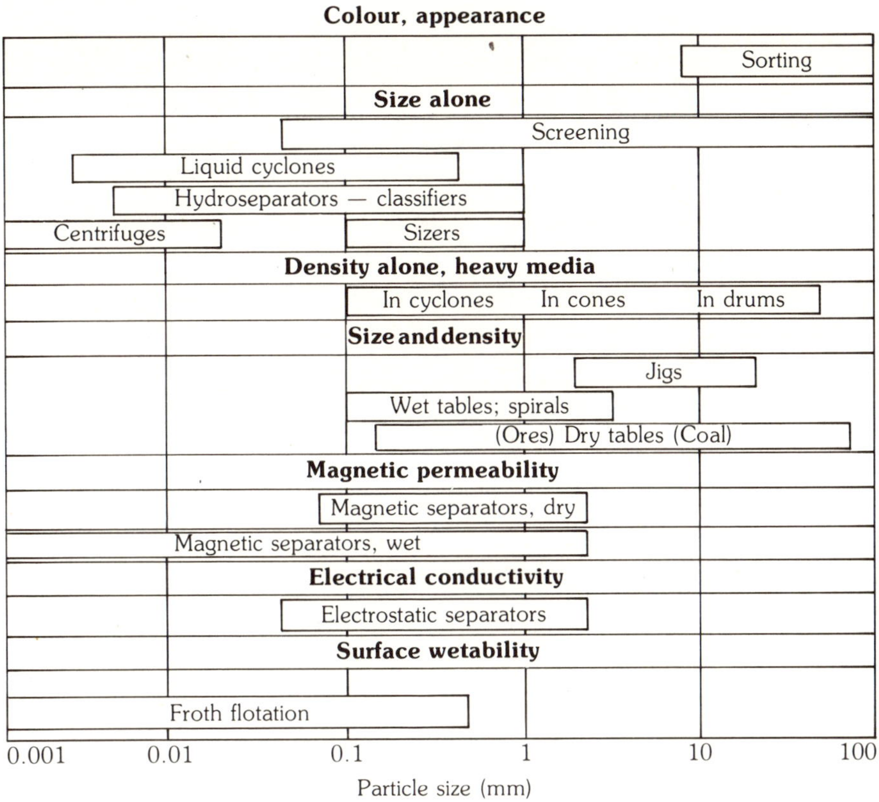

conflicting goals. Using a low magnetic field, only the purest tin cans will be attracted and separated. The yield is low, but the product grade and concentration ratio are maximal. With increasing field strength, also dirty cans, cans entangled with other material and massive iron will be collected. The yield increases, but product grade and concentration ratio go down. At the limit, heterogeneous material containing very little iron, e.g. a piece of wood with a steel nail, would also be classified as magnetic material. The yield then becomes maximal, but the product grade is extremely low.

Apart from research and development work performed by NCRR at their pilot plant in Washington DC, and demonstration plant in New Orleans, La., almost no hard data exist on the performance of individual separators. Such data are also very difficult to obtain and to generalize because of:

1. the very high cost of performing tests and thoroughly analyzing the feed material and the separated fractions,
2. the large number of variables required to characterize an individual separator,
3. the difficulty of obtaining consistent performance data on a material such as refuse.

For these reasons the following discussion of individual sorting operations remains largely qualitative.

Size Reduction

Size reduction (Fig. 9.1) is used extensively in refuse disposal as a means of densifying refuse or to accelerate composting. In sorting systems it can serve:

1. make oversized material more tractable,
2. liberate entangled material,
3. convert composite items to more homogeneous materials,
4. give a feed stream an optimum, homogenous size for a particular sorting process,
5. classify refuse components according to their brittleness, hardness, elasticity or malleability.

For all these reasons, shredding is the basic step of a large number of automated refuse classification systems.

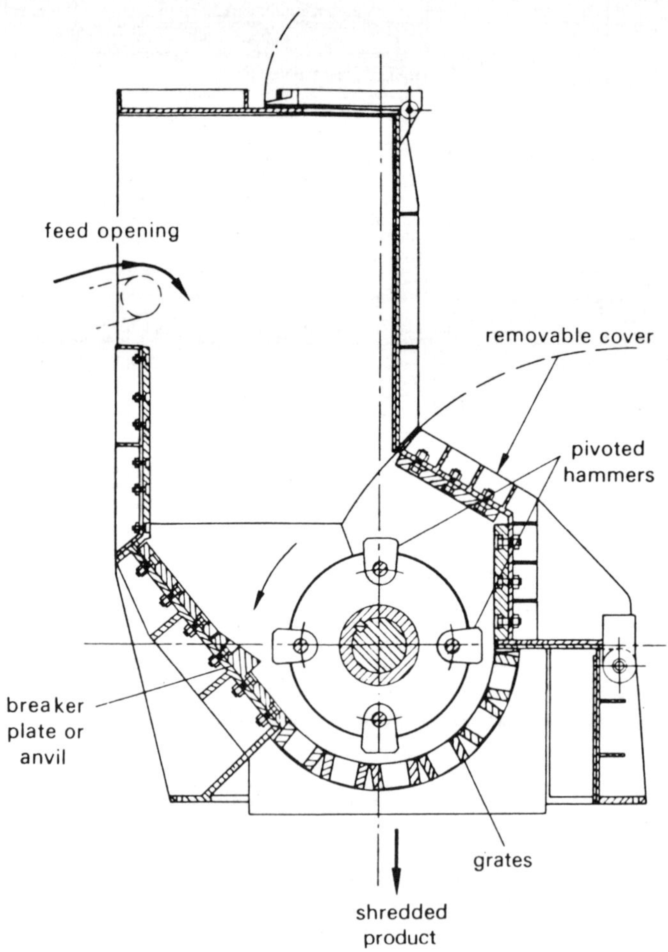

Fig. 9.1 Hammer mill. (Reproduced with permission from A. V. Bridgwater and C. J. Mumford *Waste Recycling and Pollution Control Handbook*, George Godwin/Van Nostrand Reinhold.)

In contrast a series of other systems avoid the use of indiscriminate shredding because of:

1. the high investment and operating cost (typically $10/tonne),
2. the cross-contamination of different components during shredding, e.g. the embedding of fine glass particles in paper or the fouling of paper by kitchen waste,
3. the possible occurrence of shredder explosions.

Size reduction can also be used as a selective classification method of specific feed streams:

1. Shredding of a mixture of paper and plastic film selectively tears the paper, especially when this is moist.
2. Glass can be eliminated from plastics or metals by crushing in an impact mill or a trommel with lifters.
3. Glass and aluminium can be separated by means of a roll mill. The glass is crushed and sieved off; the aluminium is flattened and remains upon the screen.

Hammer and flail mills operate on the basis of impaction and shredding. They consist of a heavy casing, with a single rotor or double rotors rotating at high speed (typically 1000 – 2000 rev./min.). The refuse is crushed by the action of hammers or flails on one hand, and of impact plates, shredding teeth and grate bars on the other. In case of overload, the feeding belt is halted automatically. The flail mill allows heavy and bulky objects to pass by without damage. Some other mills (Gondard, Tollemache) have provisions for the ballistic elimination of such objects.

Low-speed knife or disc mills have been used to reduce the size of the larger refuse components without extensive shattering of glass or cross-contamination.

The shredding of paper can be accelerated by moistening, but also by a combined treatment with heat and strong acids. In the CEA/A.D. Little process, the paper content of refuse is thus converted into a hydrophobic powdered fuel by means of a ball mill.

Screens and Sieves

Screens and sieves (Fig. 9.2) are basic classification equipment, used to:

1. protect downstream equipment against oversized and tramp material:
2. remove fines, which are either valueless, e.g. ash, or interfere with subsequent operations, e.g. by the formation of slimes;
3. prepare the feed stream to a particular sorting process, which requires material in a given size range;
4. produce a commercial product, e.g. compost, by elimination of coarser, non-composted material.

Raw or shredded refuse can be subdivided into several size fractions, having a markedly different composition:

1. *Fines* (typically below 2 cm), containing ash, dirt, earth and shattered glass.
2. *Middle fraction* (typically 1 – 2 to 5 – 10 cm), containing a large part of kitchen and garden waste, crown or other corks, small pieces of paper, plastics or metal.

Fig. 9.2 Trommel screen.

3. *Large fraction* (typically 5 – 10 to 15 – 30 cm), containing a large part of re-useable paper, plastics, cans and bottles.
4. *Oversized material*, consisting of corrugated boxes, large film, magazines, carpets and miscellaneous objects.

Several European systems therefore use screening as a first step in the classification of raw refuse.

Trommel screens, consisting of a slightly inclined, rotating cylindrical frame, covered with perforated plate, bars or wires, are most used in the screening of raw and shredded refuse. Few guidelines exist for the design of such units or its possible internals, such as scrolls to guide the flow or lifters to agitate the content.

The speed of rotation influences the extent to which light material may become airborne.

The most important operating problem with screens is their tendency to be blinded by wires, textiles, garden waste and the like. The trommel screen has only a small self-cleaning effect, in particular for the larger holes (>10 cm) and when rotation is slow. A few designs incorporate a long slender closed-loop conveyor slightly beneath the zenith line of the trommel. Flexible items are caught by this conveyor and discharged manually when they leave the trommel.

Flat, shaking, vibrating or oscillating screens do not seem to perform so well on raw or shredded refuse. Good results have still been reported at the Technical University of Aachen with a special type of sieve, the plastic bars of which are alternatively loosened and stiffened.

Ballistic Separators

Ballistic separators are used to classify refuse according to the inertia of the individual components. One by one the latter are dropped onto a rotor and projected into a classifying chamber. The greater the mass of the particle, the greater its travelling distance (Fig. 9.3).

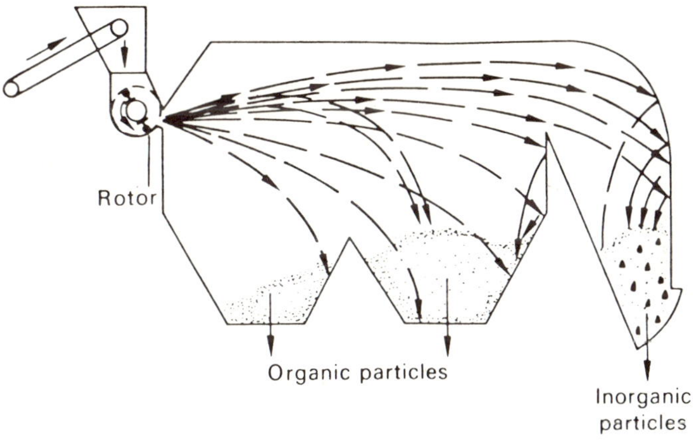

Fig. 9.3 Ballistic separator. (Reproduced with permission from A. V. Bridgwater and C. J. Mumford *Waste Recycling and Pollution Control Handbook*, George Godwin/Van Nostrand Reinhold.)

In a sorting process developed by the Franklin Institute, refuse is shredded, screened and air classified. Material over 2.5 cm in diameter is hurled away by a rotating cogwheel in an almost horizontal direction and collected by one of three consecutive bins. The last one catches dense material, such as large pieces of metal or plastics; the second one recovers larger pieces of paper and board and small pieces of metal; the first one collects paper and soft plastics. A vertical downward blast of air helps the light material to drop out first.

Ballistic separation is not frequently used in sorting systems, but both the vertically axed Tollemache (Heil in the USA) and the horizontally axed Goddard hammer mills feature ballistic separation of hard and elastic materials.

Secators

The secator (Fig. 9.4) is used to separate hard and elastic materials from softer and lighter ones. The components to be separated are dropped onto an inclined band conveyor or onto a slowly revolving drum. The heavy, hard or elastic particles bounce up and move down by gravity, whereas the light and soft material adheres to the surface and is carried away. The adhesion can be enhanced by minimizing the drop height and the inclination or sprinkling water onto the surface. The secator is only moderately successful in eliminating glass and stones from compost, part of the material to be eliminated remaining entangled. It separates glass or plastic bottles, rubber, etc., from large pieces of garden waste in the BRGM process.

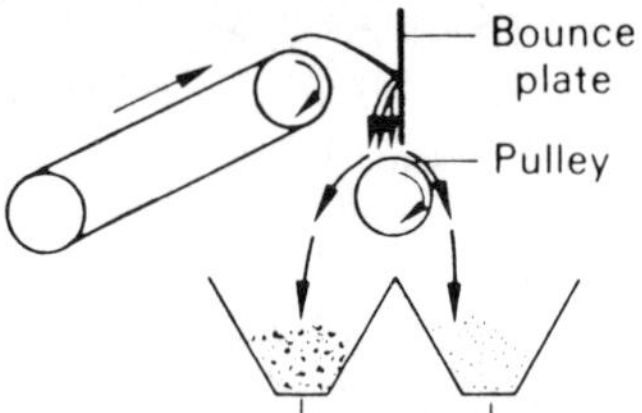

Fig. 9.4 Secator. (Reproduced with permission from A. V. Bridgwater and C. J. Mumford *Waste Recycling and Pollution Control Handbook*, George Godwin/Van Nostrand Reinhold.)

Fluidized Bed Separators

Fluidized bed separators (Fig. 9.5) have been used for separating metal from cable sheath after chopping up electric cable and for the separation of aluminium- and copper-rich fractions obtained from car shredding or incinerator ash.

The separators are based on the sink/float principle, i.e. light particles flow on top and dense particles sink to the bottom of the bed. In the DRY-FLO separator developed by Warren Spring Lab. (Stevenage, UK), a shallow inclined trough with a sintered bronze base is filled with a sized granular medium, consisting of ferrosilicon, magnetite or iron powder, which is fluidized by pressured air, distributed through the base plate. The apparent density of the bed material falls between those of the materials to be separated.

A two-mass vibratory system conveys the sink fraction upwards along the sloped bottom of the trough and then over a screen, which sieves off the entrained bed material. The floating material is carried to a discharge point situated at the other end of the separator by means of a flow in the opposite direction induced at the surface of the bed.

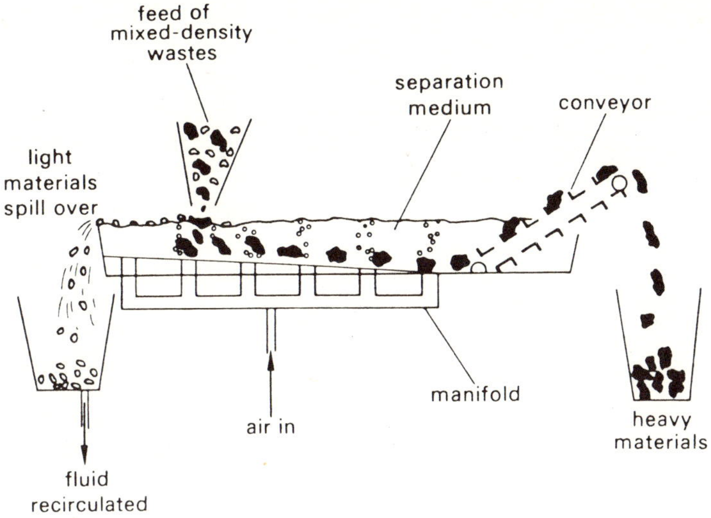

Fig. 9.5 Fluidized bed separator. (Reproduced with permission from A. V. Bridgwater and C. J. Mumford *Waste Recycling and Pollution Control Handbook*, George Godwin/ Van Nostrand Reinhold.)

The DRY-FLO separator handles material in the size range 0.2 – 10 cm; it requires a minimum difference of density of 0.2.

Stoners

Stoners are slightly inclined perforated tables with a specific movement so that dense particles are climbing faster than lighter ones. The latter are carried off downwards.

An air current, rising through the perforations, more or less fluidizes the mixture.

Air Classifiers

Air classifiers separate waste components on the basis of their falling velocity, which is determined by weight and aerodynamic properties; they are major components of almost every dry sorting system.

In the simplest type of air classifier (USBM; WSL) paper foil and plastic film are aspired from or blown off a conveyor belt.

The masking of light particles by overlaying denser material can be minimized by dropping or lancing the refuse particles into an air stream. Depending on the relative importance of gravity and drag forces, the free particles are or are not entrained by the air stream.

Fig. 9.6 shows several possible configurations of such air classifiers.

Similar results can be obtained by the use of high-pressure blowpipes. Since the momentum of the high-pressure air is rapidly depleted, it is generally possible to avoid the cleaning or recirculation of the air.

Vertical classifiers give a sharper separation than horizontal ones; on the other

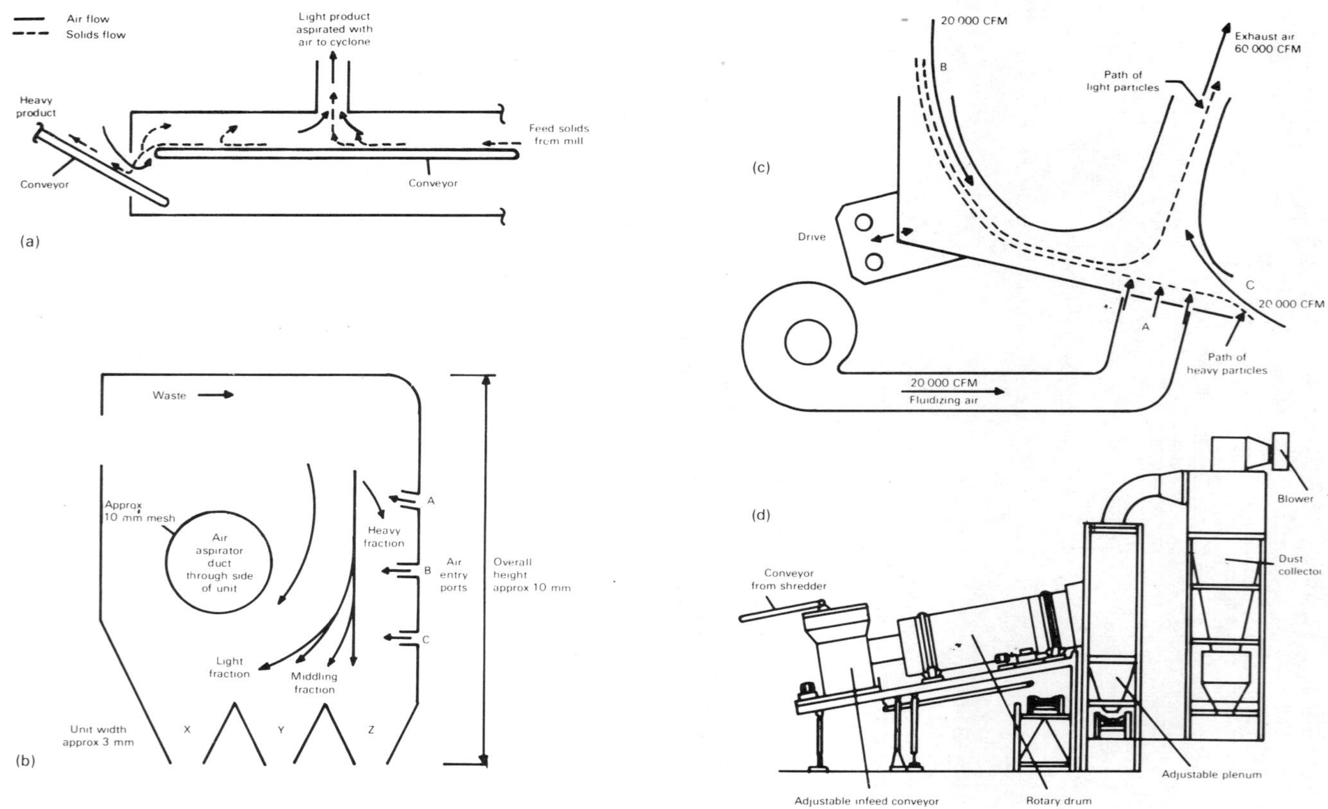

Fig. 9.6 (Reproduced form *Crest Study of Household Waste Sorting Systems: Final Report*, 1979, by permission.)

hand, their capacity is a factor of 3 – 4 lower to allow the settling of particles without excessive mutual hindrance.

A still better classification is achieved by the zig-zag air classifier (Fig. 9.7).

The lighter particles are carried over by the upward air stream, which follows an almost linear path of flow. The heavier particles settle from this stream and slide down along the walls. At each elbow, these downgoing particles are forced to pass across the rising air stream.

Air classification is often combined with inertial, ballistic or magnetic effects, either to obtain specific effects or to clean the product of a different type of separator, e.g. using an air knife.

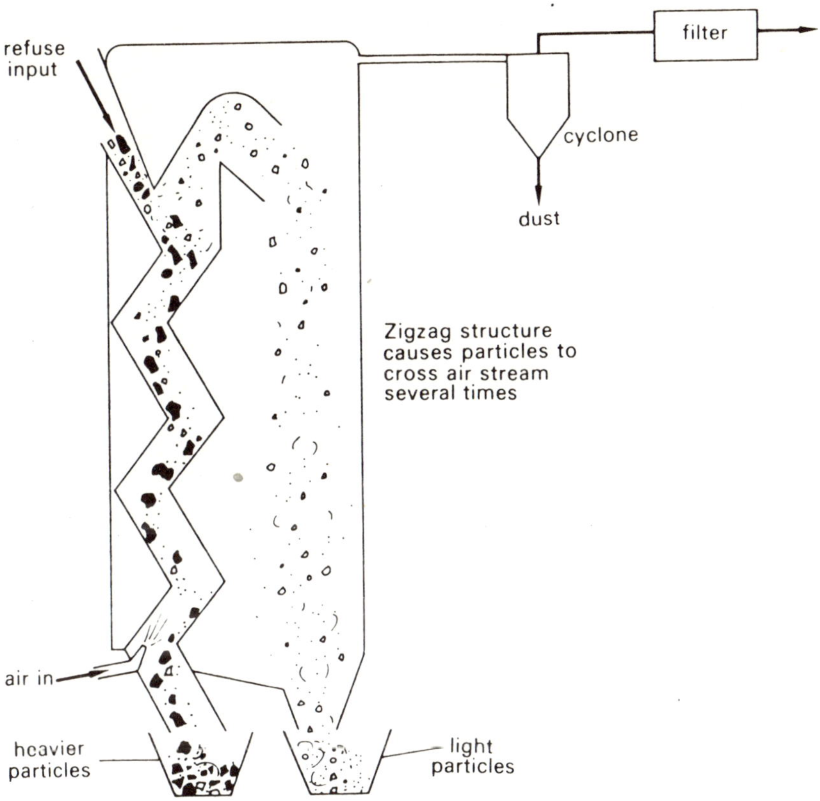

Fig. 9.7 Zig-zag air classifier. (Reproduced with permission from A. V. Bridgwater and C. J. Mumford *Waste Recycling and Pollution Control Handbook*, George Godwin/Van Nostrand Reinhold.)

Vortex Classifiers

Particles falling through an air vortex change their radial position under the combined influence of gravity and aerodynamic forces and approach their equilibrium radius in a manner depending on their density and drag coefficient. Such particles can eventually be collected at convenient points.

Vortex classification has been studied at the Massachusetts Institute of Technology (MIT). A very good agreement has been reached between experimental and computer-predicted equilibrium radii for all but very thin, lamellar particles, the drag coefficient of which is not well known.

Tentative practical applications using a confined vortex have centred on the use of cyclones and hydrocyclones. Air is used as a classifying medium when the density is lower than one, water for denser materials.

Optical Sorters

Optical sorters (Fig. 9.8) have been developed for separating colour-sorted glass from a concentrate of mixed cullet. A vibrating feeder delivers glass particles onto moving belts, on which they are isolated as individual particles. Afterwards, they fall into the sorter system, which consists of optical, electronic and ejection sections. In the optical section the falling particles are illuminated by three high-intensity lamps. Three photoelectric cells focus on a common point, to yield a 360° coverage of the passing particle. A zero condition is created by a standard background, any deviation from which is signaled to the electronic section. Particles which are brighter (or darker) than the background colour are ejected pneumatically.

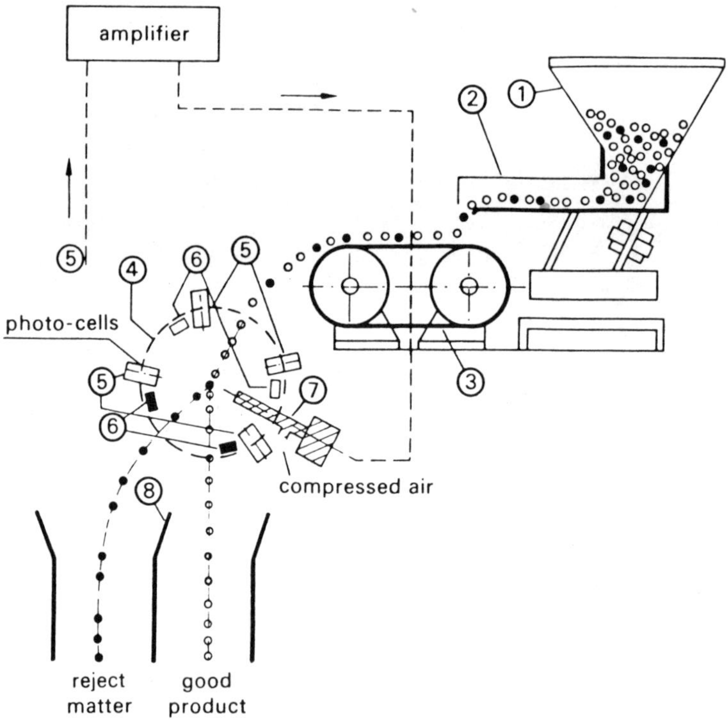

Fig. 9.8 Colour separator. Commodities to be separated are: fed from the hopper (1) by means of a vibrating chute (2) aligning themselves on a grooved belt (3). They are then project in single file into the optional box (4) and inspected by six photo-cells (four for shade and two for colour) (5) against coloured backgrounds (6). Any discoloured items are deflected from the main stream by an air jet (7) to the other side of a dividing edge for separate collection (8). (Reproduced with permission from A. V. Bridgwater and C. J. Mumford *Waste Recycling and Pollution Control Handbook*, George Godwin/Van Nostrand Reinhold.)

Dry Magnetic Separators

Dry magnetic separators are extensively used in refuse sorting, either to recover ferrous metal from raw or shredded refuse or to remove tramp iron from various streams. Both permanent and electromagnets can be used. Permanent magnets require no external power but are effective only with low speeds of the conveyor belt and low depths of the burden. Most magnets are powered by oil-cooled d.c. electric coils.

Shredded refuse contains several per cent of ferrous metal, so that a suspended magnet should be self-cleaning. The overband separator consists of a stationary electromagnet, around which a rapidly moving rubber belt is fitted to capture and carry off the material lifted from the refuse (Fig. 9.9).

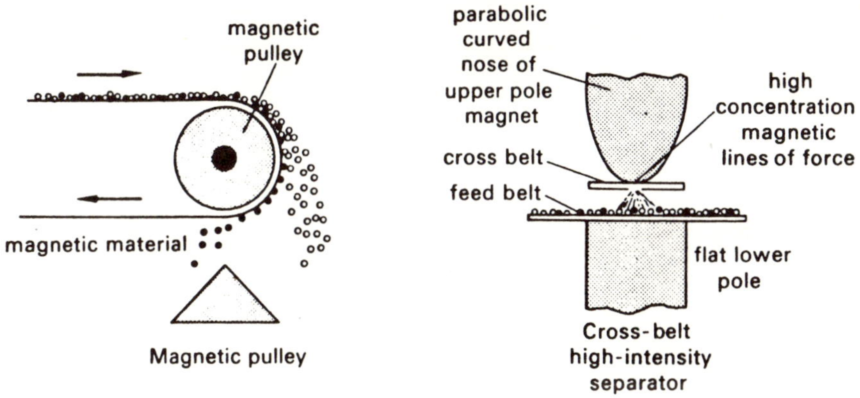

Fig. 9.9 Magnetic separators. (Reproduced with permission from A. V. Bridgwater and C. J. Mumford *Waste Recycling and Pollution Control Handbook*, George Godwin/Van Nostrand Reinhold.)

The yield of the recovery of ferrous metal depends on the strength of the magnetic field, the distance from belt to belt and the depth of the burden. On raw refuse, the yield can be as low as 25%. With shredded or sized refuse, the separator can be mounted much closer to the belt so that a yield of 50% or more can be attained.

When the speed of the refuse conveyor belt is below 1.5 m/s, a cross-flow arrangement is generally preferred. To reduce the loss of ferrous metal buried underneath non-magnetics, the overband can be installed in-line over the discharge point of the refuse. Magnetic separation is assisted by the difference in momentum.

The overband separators have several disadvantages, such as:

1. excessive wear of the belts, with typical lifetimes of the order of weeks or months, depending on the kind of duty and belt material;
2. the occurence of supplemental damage when iron gets caught between the belt and the magnets,
3. the presence of multiple bearings and pulleys in a dust-loaded atmosphere.

For heavy-duty operations, the overband can be replaced by magnetic drum separators with a hard shell of manganese steel.

The pulley separator (Fig. 9.10) is self-cleaning. It is often used for completing the removal of ferrous metal after a preliminary separation by an overband. Sometimes non-magnetics get caught between ferrous metal and the belt and report to the magnetic fraction.

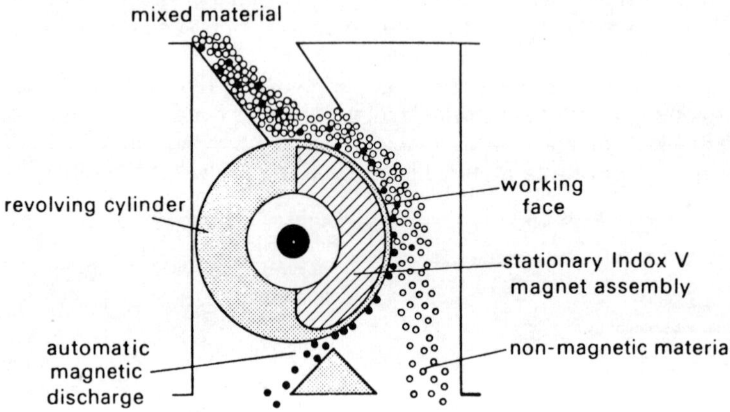

Fig. 9.10 Pulley separator. (Reproduced with permission from A. V. Bridgwater and C. J. Mumford *Waste Recycling and Pollution Control Handbook*, George Godwin/Van Nostrand Reinhold.)

The separated material still contains loose and fixed contaminants. These can be partially eliminated by:

1. using ballistic effects,
2. using air knives during the lifting, conveying or discharging of the ferrous metal,
3. exposing the lifted material to consecutive magnetic fields of alternating polarity, i.e. magnetic agitation: the attracted material is thus forced to spin and release loosely attached non-magnetics,
4. using two magnetic separators in series with intermediate shredding to liberate contaminants: shredding also increases the density of the recovered ferrous material.

Magnetic separators are also capable of separating products with different magnetic susceptibility.

Vibrating conveyors, chutes, tables or screens must be of non-magnetic material, to avoid induced magnetism.

Typically the investment cost of a single, suspended drum separator is \$22 000, with an annual operating cost of \$1000. The treatment cost is of the order of \$0.2/tonne.

Electrostatic Separators

Electrostatic separators can be used to separate conducting from non-conducting particles, or even to separate non-conductors consisting of different materials. Typical applications include moist paper/plastics, glass/aluminium, and different types of plastics.

The particles to be separated can be charged by conductive induction, ion-bombardment or frictional electrification.

The US Bureau of Mines has applied electrostatic separation to synthetic mixtures of moist paper and plastics. The mixture was supplied by a vibrating feeder onto a grounded revolving drum, after size reduction to 2.5 – 7.5 cm. A corona discharge was generated by applying a high voltage (10 – 200 kV) between the drum, whereas the paper was attracted towards the active electrode. Paper is a suitable conductor only when it has a sufficient moisture content. The same principles can be applied to the separation of glass and aluminium.

When a mixture of chips of cellulose acetate and polyethylene terephtalate (PETP), both used as a substrate for photographic film, are poured out of a container, both types of chip become charged spontaneously by friction. They can then be easily separated from each other by applying a strong electric field.

In practice it is difficult to use electrostatic separators because of the frequent occurrence of short-circuiting, the necessity to control the moisture content of paper and the presence of soil and contraries, e.g. plastic-coated paper and surface-treated plastics.

Metal Detectors

Metal detectors are used in numerous processes to detect small pieces of tramp metals. The particles to be analyzed are made to pass or fall through a magnetic field. When a metallic particle passes through the detector, an impulse is generated in the electromagnetic winding. After amplification, a deflection mechanism or an air jet is actuated by this signal, and the detected metal particle is ejected by one of these means. If the metallic content is too high, the mechanical system will reject a large part of the stream.

Eddy Current Separators

Eddy current separators have been developed to separate metal particles from non-conducting ones. When a conducting particle is subjected to a variable magnetic field, an eddy current is generated, which follows a closed circuit inside the particle. The induced current has its proper magnetic field, which is opposed to the original field. The repellent force of the two fields can be used to deflect the metal particle from the feed stream.

The action of the magnetic field on the metal particle is a complex function of conductivity, size, form and density and of the intensity and geometry of the magnetic field. Denser materials of lower conductivity and small or elongated objects are more difficult to separate. The feed stream should first of all be completely freed from ferrous particles, since the latter are retained permanently.

Eddy current separators, developed by Combustion Power Company and Occidental Research Corp., consist of a conveyor belt passing the feed material over (or under) one or more sets of linear motors. The metal particles are deflected sidewards and are collected at the side of the belt.

The Raytheon separator (Fig. 9.11) features a long ramp with permanent magnetic strips of alternating polarity embedded under an angle of 45° with the ramp axis. The feed stream slides down the ramp and the metal particles are deflected and collected separately. Raytheon has also tested a rotary-drum

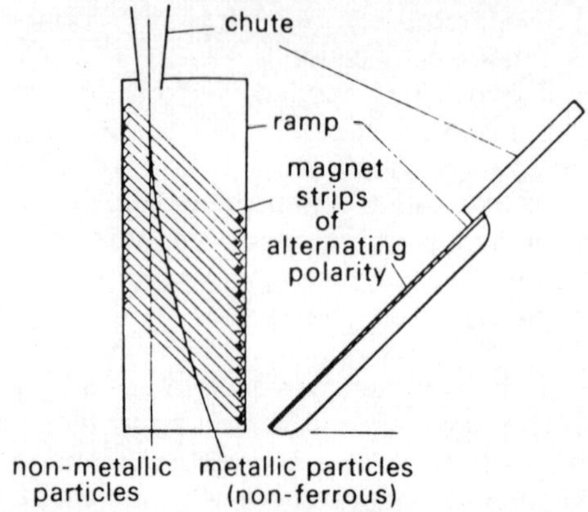

Fig. 9.11 Eddy current separator. (Reproduced with permission from A. V. Bridgwater and C. J. Mumford *Waste Recycling and Pollution Control Handbook*, George Godwin /Van Nostrand Reinhold.)

separator. A vertical slit separator with embedded magnets has been tested at the Technical University of Delft.

In actual refuse sorting, eddy current separators have been applied only in a few American plants, the aluminium content of European refuse being too low to warrant their use.

WET SORTING

Survey

Wet classification methods are extensively used in mineral dressing (Table 9.1). In refuse sorting, with the single exception of the Black-Clawson system, wet methods are used only for specific purposes, e.g.

1. to separate organics from inorganics by sink/float methods,
2. to pulp the waste paper fraction and purify the resulting pulp,
3. to classify inorganic materials, such as glass, aluminium, and nonferrous metals. Wet processing virtually eliminates explosion hazards and dust problems.

Fractions separated by wet classification are generally cleaner than those obtained by dry methods, because attached organics and dirt are largely removed. Unfortunately wet methods also give rise to a highly polluted waste water. Moreover, they are generally more expensive, in terms of both investment and operating cost.

Sink/Float Separation

A mixture can easily be separated into a light floating and a dense sinking fraction, merely by immersing it in a fluid of intermediate density in which the individual components are insoluble.

Water is a suitable fluid to separate floating organic from sinking inorganic material. Water/alcohol mixtures and brine of suitable density have been tested as a means for sorting plastics in their various types (PE, PS, PVC, etc.).

To separate materials denser than water it is also possible to use chlorinated or brominated organic liquids of suitable density. This practice has been restricted to laboratory work, because of the cost, toxicity and limited stability of these liquids (they slowly hydrolyze upon contact with water). In industrial practice 'heavy media' are used as a separating fluid. These are stable suspensions of very finely dispersed inorganic magnetic materials, such as magnetite (Fe_3O_4) or ferrosilicon (FeSi), with an apparent density of 1.25 – 3.4.

A heavy media separator consists of:

1. a sieve on which the fines are removed from the feed and the particles are sprayed with water,
2. a vessel in which the feed is divided into a floating and a sinking fraction,
3. screens on which these two fractions are separately drained and washed.

Prior to the seperation, the fines and the dust are removed, as they form slimes and increase the viscosity of the heavy media. The density of the slurry can be monitored continuously by means of a gauge, which measures the absorption loss of the radiation from a gamma-ray source.

After the separation, the sink fraction is elevated from the bottom by means of an air lift, a spiral conveyor or a rotating cage lifter. The liquor draining from the separated fractions is recycled into the separator; the wash liquor is too diluted for recirculation, but the particles in suspension are ferromagnetic and hence easy to recover by means of a wet magnetic separator.

An ingenious heavy-media separator has been built by AVCO (Lowell, Mass.) for NASA. The separating vessel is installed between the two poles of an electromagnet, which generates a magnetic field with a constant gradient in the vertical direction. The apparent density of the magnetic heavy medium can be varied continuously between 1 and 25 by adjusting the intensity of the magnetic field. Non-magnetic objects remain unaffected and keep their normal physical density. They can be separated by floating one after another, at an appropriate apparent density.

The apparent density of a metal is somewhat affected by its shape and size. A more serious problem is the possible presence of bimetallic objects, objects with entrapped air or parts containing iron bolts or clips. The latter will be extremely difficult to float. The medium used contains colloids of a size of about 1 μm suspended in kerosene and stabilized by surface active agent. The 'ferrofluid' is eventually recovered by washing with a voltatile solvent, which is later eliminated by evaporation.

Wet Classifiers

Wet classifiers are based on the difference in settling rates of dense and light particles of a constant dimension or between coarse and fine particles of a constant density. The driving force may be either gravity or centrifugal force, as in hydrocyclones and centrifuges.

The settled material can be moved from the bottom of a decanter by means of a flighted chain belt, a screw conveyor or a rake.

Elutriators

Elutriators can be used to classify incinerator residue, car scrap, chopped electric cable or colliery wastes. In an elutriator, the particles settle against a rising current (as in air classification), which increases the effective density of the liquid. The drag force on the particles is a complex function of size and form; it can be varied by adjusting the pumping rate of the classifying medium or the cross-section of the separating chamber.

The WEMCO RC-elutriator is used to separate metals and glass from suitably prepared refuse, i.e. after shredding to a nominal size of 4 – 5 cm, removal of most organic material by air classification, magnetic separation of ferrous metal and elimination of both fine and oversized material by screening.

Materials of similar density can be separated by elutriation in case they have different form factors. Glass particles, for example, have a smooth form and sink in a rising current, whereas aluminium particles, distorted by shredding, offer a higher resistance to flow and are swept upwards.

In the Bowerman-classifier (Eric County Prototype) size-classified solid waste is delivered into a hopper, in which it is wetted. The material then is dumped into the separation tank by an inclined conveyor. The settling particles are classified by a horizontal, laminar stream of water. The particles are collected in a series of consecutive baskets that can be lifted out for evaluation and recovery. The non-settling compounds are collected on a travelling screen situated at the end of the classifier, and discharged onto a conveyor belt.

The Hydrocyclone

A hydrocyclone is a vessel with a cylindrical-conical shape and provided with a tangential inlet. The entering slurry whirls down along the wall of the vessel. Because of centrifugal forces, heavier particles are concentrated near the wall, and are discharged through the apex. The rest of the slurry moves to the centre of the vortex, and leaves the hydrocyclone through the overflow.

Hydrocyclones are built in a range of sizes from about 1 cm to 1 m in diameter. They are effective in separating solids from about 5 μm to 150 mesh.

The quality of the separation depends on the geometrical properties of the hydrocyclone, the size and density of the particles and the density and viscosity of the liquid medium. The high shearing forces can be used advantageously in the processing of thixotropic fluids, but may destroy flocculated material.

DSM have developed a heavy-media cyclone for classifying car scrap.

The Centrifuge

A centrifuge can also be used for classifying solids. Since the entire body of liquid rotates at an almost uniform velocity, the shear forces between successive concentric layers of fluid are very small. Therefore the agglomeration of particles, which is essential to the thickening of dilute slurries, does not take place in a bowl centrifuge.

In nozzle-bowl machines, the feed enters near the axis of rotation and spreads radially between a number of closely spaced, perforated discs (Fig. 9.12). The feed splits into a dense fraction, which is collected by a discharge nozzle at the periphery of the bowl, and into a light fraction, which discharges near the axis of rotation. Sharp classifications are possible for particles below 10 μm. Larger particles would cause excessive erosion on the bowl.

The solid bowl centrifuge is generally a horizontal, cylindrical-conical vessel with the feed entrance near the axis of rotation. It is also capable of handling

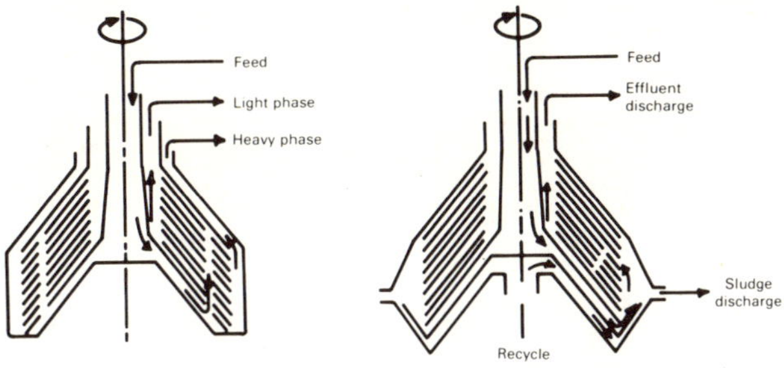

Fig. 9.12 Nozzle-bowl centrifuge. (From *Chemical Engineering Handbook*, by Perry and Chilton. Copyright © McGraw-Hill. Used with the permission of McGraw-Hill Book Company.)

coarser solids, in the 10—30 μm range, but not of complete clarification. Hence, it is sometimes used as a pretreatment before a final clarification in the nozzle-bowl centrifuge.

Jigs

Jigging allows classification of a material supported on a submerged screen (Fig. 9.13). Water is forced through the material, either by moving the screen up and down or by means of an alternating upward and downward flow (frequency is typically 2.5 stroke/s, with a stroke length of 0.5 – 5 cm). After a while the dense material can be drawn off at the bottom and the light material from the top of the bed. A third fraction consists of the fine and dense material that passed through the screen.

To obtain an efficient separation, a sufficient layer of dense material should be present. Jigging is gradually being replaced by shaking tables (below 10 mesh), sink-float methods (above 10 mesh) or air-operated jigs (stoners).

Tables

Tabling (fig. 9.14) is a process for concentrating ores and coal, acting on the basis of a difference in specific gravity, but also in the size and shape of the particles. A suspension of the mixture is fed across the slightly inclined, differentially shaken, riffled plane surface. The table motion consists of a slow forward stroke, followed by a quick return stroke, so that the heavy material on the table moves forward faster than the light material. Water is washed over the table in a direction perpendicular to that of the tabled motion, and carries along the lighter particles, whereas the denser particles settle between the ridges. The latter consist of wooden lathes or rubber strips, with a height of 6 – 25 mm. Several fractions (light, middlings, dense) can be collected by installing launders at the low end of the table. Tabling is less suitable for fine particles or when the difference in density is small.

Humphreys Spirals

Humphreys spirals consist of a cast-iron trough of curved cross-section, wound

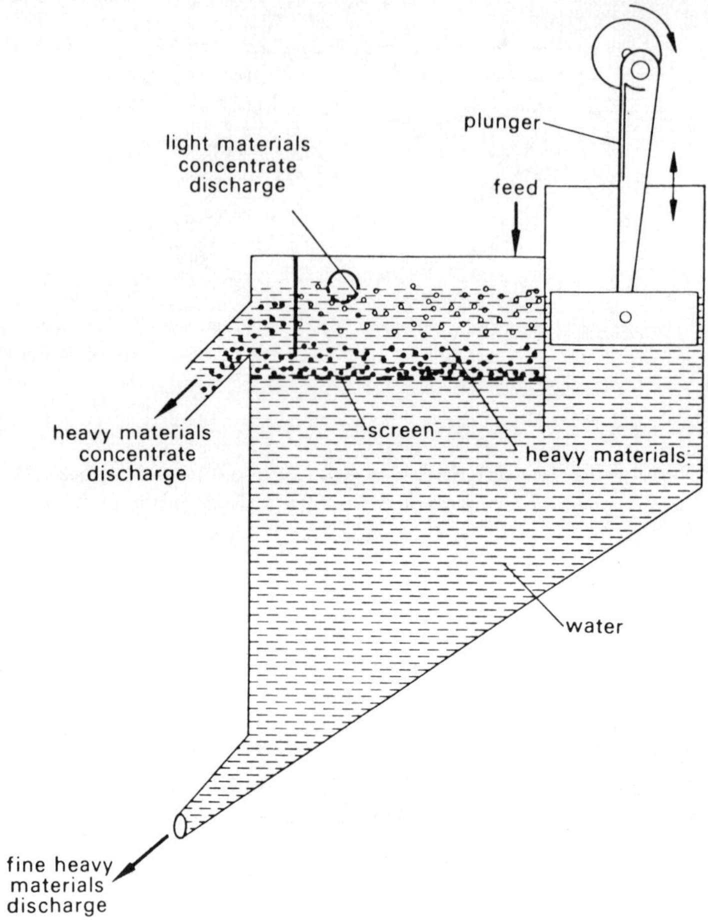

Fig. 9.13 Jig. (Reproduced with permission from A. V. Bridgwater and C. J. Mumford *Waste Recycling and Pollution Control Handbook*, George Godwin/Van Nostrand Reinhold.)

in a spiral with an outer diameter of about 0.6 m. The mixed feed is introduced at the top of the spiral. Heavy material keeps to the bottom of the trough and can be discharged through a number of ports along this bottom. Lighter material rides up the outer side and is discharged separately at the end of the spiral. Spirals can treat dense materials in the range of 10 – 200 mesh as long as the densities of the materials to be separated are significantly different.

Flotation

At the present time, flotation is probably the most important method for processing complex ores. The ores are first finely ground to liberate individual materials and slurried to a pulp with 15 – 35 wt % of solids. Chemicals are added which influence the surface properties of the materials to be separated. Adsorption of collectors makes particles water repellent (hydrophilic). Then a froth layer is created, either by mechanical means or by air agitation. The

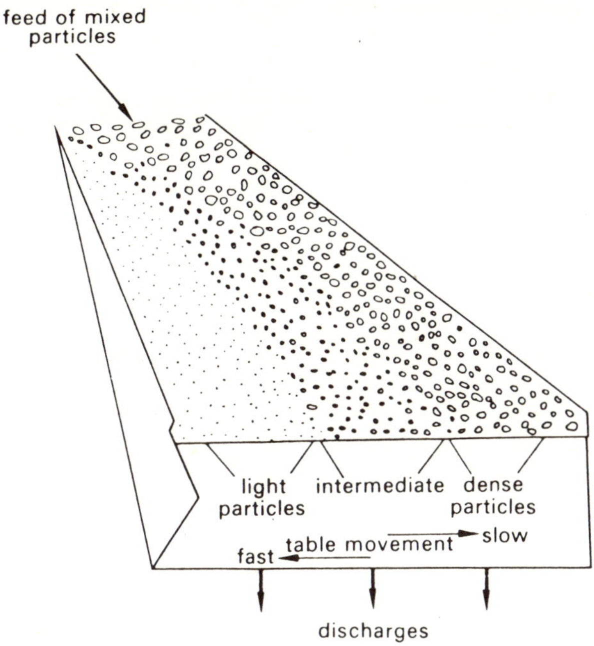

Fig. 9.14 Table. (Reproduced with permission from A. V. Bridgwater and C. J. Mumford *Waste Recycling and Pollution Control Handbook*, George Godwin/Van Nostrand Reinhold.)

hydrophobic minerals collect in the froth and are scraped off, whereas the hydrophilic material is retained in the pulp.

A flotation unit consists of a series of consecutive cells, each of which is a simple, stirred tank, often with a distributor for low pressure air beneath the agitator. The separation efficiency depends on the selectivity of the added chemicals, the pH value and temperature.

Flotation has been studied as a method of recovering metals from incinerator residue (USBM) and of finely dispersed glass from a glass rich fraction (Occidental Research, WSL, NCRR).

The separation of oils, fat and proteins from various waste waters, the recovery of cellulose fibre from the white waters of the paper industry and the elimination of printing ink and waxes from waste paper is also based on flotation. In these cases, however, the physicochemical background and especially the flotation plant involved is different.

Flotation can be applied to all liberated particles that can be lifted by gas bubbles (5 – 800 μm). Flotation costs typically a few \$/tonne, depending on the reagent consumption and the complexity of the process.

INTEGRATED SORTING SYSTEMS

Introduction

The oldest integrated sorting system, developed by SORAIN-Cecchini in Rome, has been in continuous commercial operation since 1964. The tipping fee paid by the City of Rome is lower than the normal cost of refuse disposal in a large City. If this example has not been copied yet elsewhere, it is mainly because the market conditions for secondary materials are very favourable in Italy and also because a large part of the involved technology has long remained mysterious and undisclosed.

The concept of regarding every waste as a potential source of raw materials has been introduced in the USA by the former US Bureau of Solid Waste Management (now part of EPA) and stimulated by financial means provided by the Resource Recovery Act of 1970. Since then, refuse sorting has been regarded as a means of saving natural resources, reducing environmental pollution and, in many countries, of diminishing dependence on imported raw materials. Governments in all industrialized countries consequently have stimulated research and development in resource recovery.

Thus the development of integrated sorting systems has been more a consequence of fundamental economical, political and even philosophical considerations than of a pressing demand from industry for secondary raw materials. Probably American industry is even less adapted to the recycling of wastepaper, cullet or cans than is European industry. This is reflected in the comparative simplicity of most American refuse-sorting plants, compared to the European developments. Many American plants aim at the production of waste- or refuse-derived-fuel (WDF or RDF), whereas most processes in Europe instead recover the paper content of refuse as a fibre stock. Still, the larger rate of generation and the higher grade of American refuse holds the promise of some economies of scale and of recovering materials, such as aluminium or other nonferrous metals, which are not present in sufficient amounts in European refuse.

Japanese efforts in resource recovery from wastes have been reviewed elsewhere[10].

Effects of Collection

The collection method and frequency has a marked influence on the quality of the refuse to be sorted:

1. The frequency of collection determines the freshness or putrescence of the refuse, with potential concomitant problems of odour hindrance, flies and moisture transfer. Frequent collection is particularly desirable in the countries of Southern Europe, where the refuse has a high organic content.
2. Where refuse is collected in paper or plastic bags, provision should be made to open the bags and liberate their contents prior to or during the initial stages of sorting.
3. In compression vehicles, the mixing and contamination of refuse components are minimal. Screw compacting systems open most refuse bags and cause a more intensive mixing. The tumbling action of rotary drum collection vehicles opens most bags and causes extensive moisture transfer and attrition.

Frequent refuse collection with a compression vehicle seems optimal with regard to central sorting.

Effects of Refuse Composition

Several factors have a major influence on refuse composition and hence on the possibilities of using central sorting as a disposal method:

1. The use of solid fuel for domestic heating introduces a strong seasonal variation of refuse composition, density, size distribution, moisture content and intrinsic value. The presence of ash may lead to increased wear and dust problems and decrease the value of products, such as wastepaper, fodder or compost, by contamination with ash.
2. The content of organic material has a determining influence on the moisture content and hence on the stability and putrescibility of refuse. Although organic material can be valorized as a fodder, compost or fuel (after drying) it decreases the recycling potential of refuse.
3. The paper content of refuse, which is high in Scandinavian countries, the USA and Canada, intermediate in W. Europe, and low in S. Europe and developing countries, determines the possible yield of WDF or wastepaper, one of which is generally the main product from refuse sorting, both in terms of tonnage and of revenues.
4. The type, materials and extent of packaging is a most important factor, since packaging materials are the most valuable components of refuse. The consumption pattern of these materials varies with time and location, and this influences the yield and grade of the fractions recovered from refuse. The introduction (or prohibition) of non-returnable glass containers, for example, determines the amount of cullet that can be extracted from refuse. PVC bottles are to be found in some countries only, e.g. France and Belgium, so that their recovery may be warranted there and not in other countries. Heterogeneous packaging materials, such as tin cans with an aluminium lid or plastic-laminated paper, may create problems with regard to the use of the sorted fractions.

The recycling of materials from waste by other means, such as separate collection, generally does not have a decisive influence on refuse composition. Yet it may compromise central sorting in that it generates secondary materials of a more consistent and better quality and thus closes the outlets for the fractions obtained by central sorting.

Structure of Sorting Systems

The next sections survey a series of European and American sorting processes. Practically all systems are based on the following four primary operations: size-reduction, screening, air classification and magnetic separation. These can be combined in various sequences and each operation can use different types of equipment. Although almost each system developer has announced specific choices with regard to the operations and equipment used in his process, its final design will no doubt be profoundly influenced by the type of refuse to be processed and the kind of products to be obtained.

Size reduction of raw refuse is a vital step in most systems. Fastly rotating hammer mills or flail mills are most commonly used, but slowly rotating disc mills have been preferred in two systems. Each mill has its specific merits and reliable and comparable experience factors on shredding performance, power consumption, wear, unscheduled downtime, and safety and environmental records are almost completely lacking. Other important factors in shredder selection are the amount and nature of bulky wastes to be processed, the required size range of the shredded product, the presence or absence of abrasive material (glass, ash), the possible recovery of glass in the form of discrete particles, and the degree of product contamination caused by shredding.

A preliminary sizing of raw refuse by means of a trommel or other type of screen allows to eliminate material of no interest, such as ash and organics, at an early stage of the process and to divert part or all of the refuse from the shredder, which is an expensive piece of equipment to own and operate. This advantage should be weighed against disadvantages, such as the possible presence of elongated objects in the undersized fraction or the potential loss of oversized or bulky materials, such as large plastic film or telephone directories, recovered more easily after shredding.

Screening is also required to eliminate fines from the light fraction. The optimal location(s) for this step is influenced by the amount of fines present in the refuse generated in the process.

Air classification in recent systems is conducted either in a zig-zag classifier or in a rotary drum, which is a more sophisticated and improved version of the horizontal classifier. The zig-zag classifier is capable of excellent performance, but requires more even feeding and careful sizing to eliminate objects that could cause obstruction.

Magnetic separators have been placed in almost every conceivable location of the various systems. Used ahead of the system it can prevent the entrance of massive pieces of iron in the shredder. On the other hand, better performance is possible using a sized feed stream, in which the cans have been cleaned by preliminary shredding and air classification.

The SORAIN Plants in Rome

The Societa Riutilizzazione Agricola Industriale is responsible for the collection and disposal of the largest part of the refuse of the City of Rome. Since 1966, the latter is systematically sorted into saleable materials, such as paper pulp, animal food, ferrous metal and compost. Bulky combustible refuse and non-reclaimable fine material is incinerated and thus serves to generate the heat for sterilization of animal fodder and drying of paper pulp.

Refuse in Rome is collected daily (since February 1980 on a 6 day/week schedule) without much compaction, and delivered to one of three SORAIN sorting plants (capacity 600 tonne/day), the primary separation modules of which operate on a 2 shift/day basis.

The refuse, collected in polyethylene bags, is levelled out hydraulically and passes over a bag splitter which liberates the contents of bags and boxes. The material is sized in two consecutive trommels, the first yielding an oversize fraction and a fraction below 20 cm. The latter is separated in a second trommel into:

1. fine organic material and dirt (below 4 cm), to be composted,
2. organic material, some cans and glass (between 4 and 9 cm),
3. the oversize (between 9 and 20 cm) consisting mainly of paper and plastics.

After removal of magnetic metals, the oversize fraction is air classified to eliminate residual glass, wood, etc. which join the material of the 4—9 cm fraction.

The paper and plastics are separated by an undisclosed process based on the selective shredding of paper. The plastic film, which in Rome consists almost exclusively of polyolefins, is granulated and sold.

The coarse organic material, mainly vegetable and fruit waste, is transformed into animal fodder. The process involves:

1. elimination of fine dirt by washing,
2. elimination of glass, stones and metals by sink/float separation,
3. washing and pressing to remove excess water,
4. sterilization at 130 °C yielding a paste-like product,
5. dehydration,
6. drying and elimination of fines, paper and plastics by means of a fluidized bed separator,
7. grinding, pelletizing, packaging and storage.

The ferrous metal is purified and detinned in a rotary kiln by sweating at 450 – 600 °C. The cleaned scrap is baled and sold to mini-steel plants producing reinforcing bars.

Until now the SORAIN-Cechini plant has the longest and best record of consistently producing and selling its products, generally on the basis of annually renewable contracts.

Other European Systems

The Dutch TNO system has been developed since 1970 at successively larger scales of 0.1 tonne/h (1972), 1 tonne/h (1973), 5 tonne/h (1975) and 15 tonne/h (1976). In 1973, collaboration with two industrial partners, Recycling BV and Esmil BV, started. The final demonstration plant has been installed at an existing refuse shredding plant in Haarlem.

The system is composed of (1) a bag opener, (2) a magnetic separator, (3) a hammer mill, (4) a zig-zag air classifier, (5) a rotating screen, (6) a humidifier, (7) a second zig-zag air classifier, (8) a drier, and (9) a paper baler. The separation of paper and plastic film is based on the selective moistening of paper, which becomes denser and is separated during a second air classification step.

The waste paper obtained has been evaluated as a raw material for producing board and schrenz, i.e. the fluting paper used in corrugated paper. The mechanical properties are excellent, possibly due to the presence of good quality kraft wrapping paper in the mixed product. The recovered cans are suitable for detinning. The fine fraction can be either composted or added to the clay in the manufacturing of large insulating bricks.

A 40 tonne/h plant is under construction in Austria.

The Swedish Fläkt System has been tested in a 5 tonne/h plant, constructed in 1975 at the site of the Hogdalen incinerator near Stockholm. A 25 tonne/h demonstration plant is in commission at Wijster (Holland). The system consists of the following series of steps.

1. Coarse shredding of refuse in a double-rotor flail mill without bottom grate. The mill opens the bags, coarsely shreds the paper and plastics, shatters the glass and allows large and heavy objects to pass through without reduction in size.
2. Primary trommeling of the shredded refuse. Originally the oversize material (> 20 cm) was intended for recirculation to the shredder. Now it is considered a means of extracting large plastic film.
3. A zig-zag air classifier separates the light from the dense material. The latter is discarded after magnetic extraction of ferrous material.

4. The light fraction is treated in a second shredder, followed by a second trommel. The trommel yields three fractions. The fines (<1.5 cm) are suitable for composting, the coarse fraction (>10 cm) concentrates plastic film and the middle fraction consists largely of paper (85%), plastics and textiles (10%).
5. The paper-rich fraction is flash dried in a two-stage pneumatic drier, to dry and hygienize the paper stock. Some dense, foreign material drops out in the first stage. Thermoplastic film shrinks to crisped particles, which can be eliminated in a consecutive air classification step. Adhering dirt is set free when the paper passes through the fans of the drier plant.
6. The dried paper is air classified in a second zig-zag classifier. The top product consists of very pure newsprint (3% impurities). The dense fraction consists of denser packaging paper, together with almost all contraries, such as plastic or aluminium laminates, textiles, etc. (30% impurities).

In W. Germany, Prof. Hoberg and collaborators (Technical University of Aachen) have since 1972 developed a comprehensive sorting system, which combines dry and wet sorting methods. Tests took place at pilot scale (1 – 1.5 tonne/h).

In this system the moist, organic refuse is eliminated at an early stage to avoid contamination and moistening of paper and diminish the capacity required downstream. The material below 20 mm (ash and dust) and 40 mm (organic material) is sieved off on a special type of vibrating sieve. The sieving surface is alternately stretching and loosening; the resulting 'trampoline' effect improves the sieving action and avoids blinding.

The coarse fraction is shredded in a slowly rotating disc mill, which, acting by compression and shear, fragments the bottles without extensive splintering. The disc mill does not wind textiles or nylon stockings or give rise to explosions. The shredded material is separated in three fractions by means of a zig-zag air classifier. The first consists of recuperable paper and plastics; the second of organics, together with moist, dirty or laminated paper; the third concentrates the valuable dense material.

This last fraction is cleaned by elutriation. The coarse organics are scraped off and separated on an inclined screen. The water can be recycled after settling without requiring further treatment, since it is being replaced by fresh water to make up for the losses.

The sinking fraction is eliminated from the elutriator and dewatered by means of a screw conveyor. An overband recovers fairly clean can stock and the ceramics, glass and nonferrous metal are segregated by means of heavy media. The glass fragments in the size range of 5 – 20 mm are colour sorted by means of Sortex optical sorters.

The R-80 Babcock Krauss-Maffei system has been developed since 1974. After test work at 5 tonne/h scale in Munich, a plant with a capacity of 10 tonne/h has been built in Landskrona (Sweden). The system combines ferrous removal, size reduction by means of a disc mill, sieving and zig-zag air classification.

A gigantic air classifier combining ballistic and aerodynamic effects was constructed in 1976 at W. Berlin by the Karl Fischer company, to classify a middle fraction of screened refuse and evaluate the light material obtained as a WDF in a cyclonic combustion furnace. The experiments were halted by a fire, which destroyed the plant.

Developments in refuse sorting in the UK are mainly based on research and development work performed at Warren Spring Laboratory (WSL) at Stevenage. A plant with a capacity of 10 tonne/h has been put to work in Doncaster on the basis of a WSL design. After experiments on Doncaster refuse, a process scheme based on trommel sizing and further processing to yield a WDF, ferrous metal and glass was worked out.

The Tyne and Wear plant at Byker is a single stream plant with nominal capacity of 30 tonne/h. The refuse is first shredded, because of the wide variations in the size distribution of the raw refuse. The plant produces WDF and ferrous metal. It features a Newell Dunford Engineering air classifier prototype, which basically consists of an almost horizontal rotating cone, with ballistic feeding of refuse at the wide end of the cone. Light material is carried out of this end by means of a spiralling air flow, entering at the narrow end. The tumbling action of the cone also cleans the dense materials.

The French Bureau de Recherches Géologiques et Minières (BRGM) has since 1975 also developed a sorting system for refuse. A pilot plant of capacity about 2 tonne/h has been tested on refuse collected form bins in the City of Orleans.

The refuse is gently sized in a large, slowly rotating trommel, with two screening sections, a first with circular apertures of 15 cm, a second with rectangular apertures of about 15 × 70 cm, which collect newspapers and magazines. The oversized material, consisting of corrugated boxes, old carpets, large textiles and films, can be hand sorted easily.

The fraction of −15 cm is further processed in the following sequence of steps:

1. removal of cans by an overband magnet,
2. sieving off the fines (<15 mm) for landfill or composting and middlings (15 – 50 mm) for production of animal fodder or compost,
3. air classification in a horizontal classifier with series of high pressure air knives and a partioning roll. The entrained fraction consists of light paper and plastics. The dense material then falls onto a wetted conveyor belt. Soft materials, such as indoor plants or large garden waste, are carried away and hard or elastic materials, such as glass and plastic bottles, or rubber, bounce back. The glass bottles then are selectively broken in a fastly rotating drum with internal lifters. Remaining plastic material is separated on air tables.

The BRGM system is being marketed actively at present, but no plant has yet been sold.

The Sociéte d'Etudes et d'Ingéniérie (SOCETING) has revamped a refuse shredding and composting plant at Laval (France). The shredded refuse after separation of ferrous metal and fines is further milled, purified and pelletized. The pellets are sent to a local district heating scheme.

The US Bureau of Mines Sorting System

The US Bureau of Mines has pioneered the recovery of wastes since 1966. It developed sorting systems for household refuse and incinerator ash and tested different separation processes. In 1973 the major equipment for a continuous 5 tonne/h refuse sorting plant was installed and later tested at College Park, Md.

The simplified flow-sheet of a typical USBM refuse sorting system is shown in Fig. 9.15. It consists of:

1. coarse shredding by means of a flail mill,

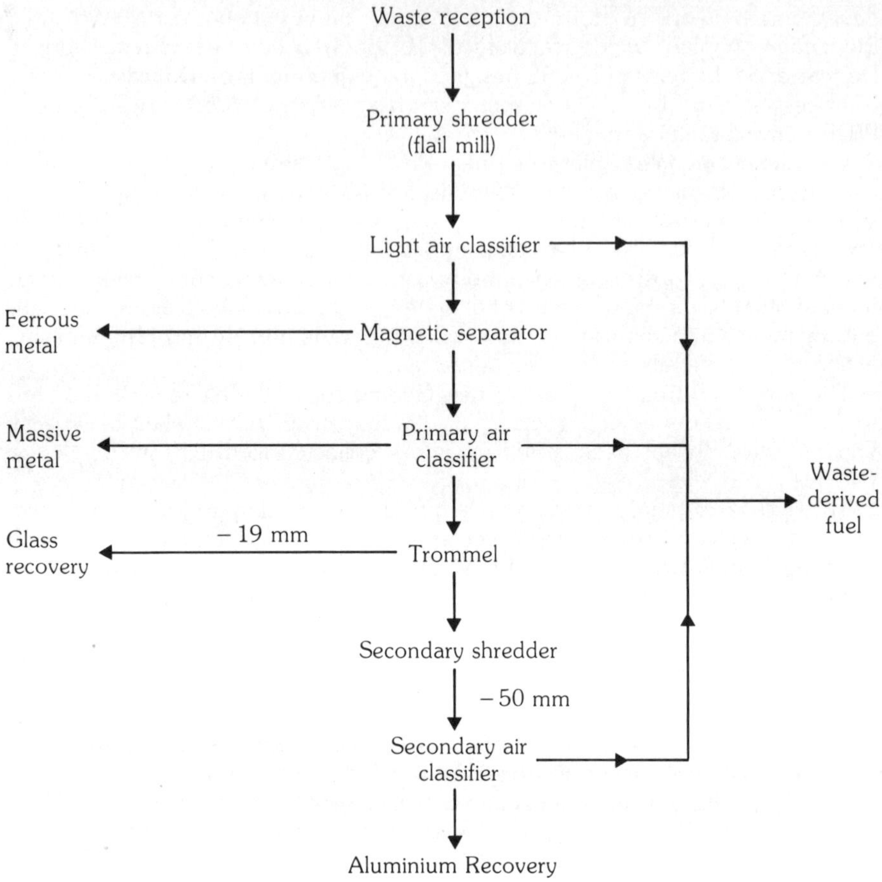

Fig. 9.15 USBM refuse-sorting system.

2. separation of light paper and plastics from the shredder discharge by means of a suction hood, followed by a cyclone, a baghouse, dust collector and an extraction fan,
3. magnetic separation of ferrous metal from the remaining stream of heavies by means of an overband,
4. horizontal air classification yielding three product fractions: a heavy fraction, which is further separated in massive ferrous and nonferrous product; a middling fraction, which concentrates heavy organics, glass, ceramics and aluminium; and a light fraction, consisting mainly of paper and plastics,
5. screening of middlings at 19 mm in a trommel: the undersize (< 19 mm) forms the feed of the glass recovery unit, the oversize is shredded in a knife mill to yield a -50 mm fraction, which is air classified to yield a light, combustible fraction and a denser, aluminium fraction, further treated in the nonferrous recovery unit.

The *light fractions* obtained in successive air classifiers can be combined to yield a paper and plastics or a fuel. The *ferrous fraction* is separated into three fractions by means of a horizontal air classifier. The heavies consist of massive pieces of iron or steel. The lights consist of loose contaminants. The middlings, mainly tin cans, are cleaned by shredding and magnetic recovery of the can fragments.

The *glass recovery section* treats roughly 25% of the total refuse stream. A battery of jigs yields four different fractions:

1. floating organic material which can be used as a fuel after dewatering,
2. the top fraction of settled solids, consisting of glass, ceramics and aluminium,
3. the bottom fraction, consisting of heavy nonferrous metal,
4. the siftings of the jig's screen, mainly fine glass.

After jigging, the glass-containing fractions 2 and 4 are combined and treated in a roll mill. The aluminium is flattened and can be sieved off easily. The glass is purified by froth flotation.

A Raytheon eddy current separator treats the nonferrous material and yields a metal concentrate, a heavy combustible stream containing little metal, and a stream of middlings. The latter can be separated by means of an electrostatic separator after drying the feed stream.

The USBM has also tested various other methods of separation or re-use of waste components, e.g.

1. separation of paper and plastics by electrostatic techniques,
2. classification of polyethylene, polystyrene and polyvinyl chloride by sink/float and elutriation techniques,
3. pyrolysis of PVC at 350 °C as a means of local HCl generation,
4. separation of aluminium from glass by selective milling and from mixed nonferrous metal by means of heavy media: the remaining nonferrous metal has been processed by vacuum and flux refining and by formation of intermetallic compounds,
5. methods were studied to re-use ferrous metal, contaminated with copper or tin,
6. glass has been evaluated as a raw material for containers, bricks, tile, glass wool and reflecting beads. Optical colour sorting was also treated.

The pioneering work of USBM has inspired many other sorting systems. Raytheon designed a Resource Recovery plant for Monroe County, NY, on the basis of the USBM System (Capacity 2000 tonne/day). The main new feature of the plant is the use of a rotating drum air classifier with axial air stream for the primary separation of light paper and plastics. Enadimsa installed a USBM pilot plant in Madrid (1973). After preliminary testing it found that the system was ill-adapted to Spanish refuse and extensive modifications were required. A demonstration plant using the Enadimsa version of the USBM process was erected in 1977 in Tournan en Brie (near Paris) by the French firm SOCEA.

The Black Clawson 'Hydrasposal' or 'Fibreclaim' Process

The Black Clawson company, a traditional supplier of equipment to the pulp and paper industry, developed a proprietary technology for the recovery of materials from refuse. The process is based on the pulping of refuse, followed by the purification of the pulp and the classification of non-pulpable materials. A simplified flowsheet is given in Fig. 9.16.

In the hydrapulper, refuse is converted to a 3 – 4% slurry containing paper fibre, organics and small pieces of glass, stone and metal. The pulper consists of

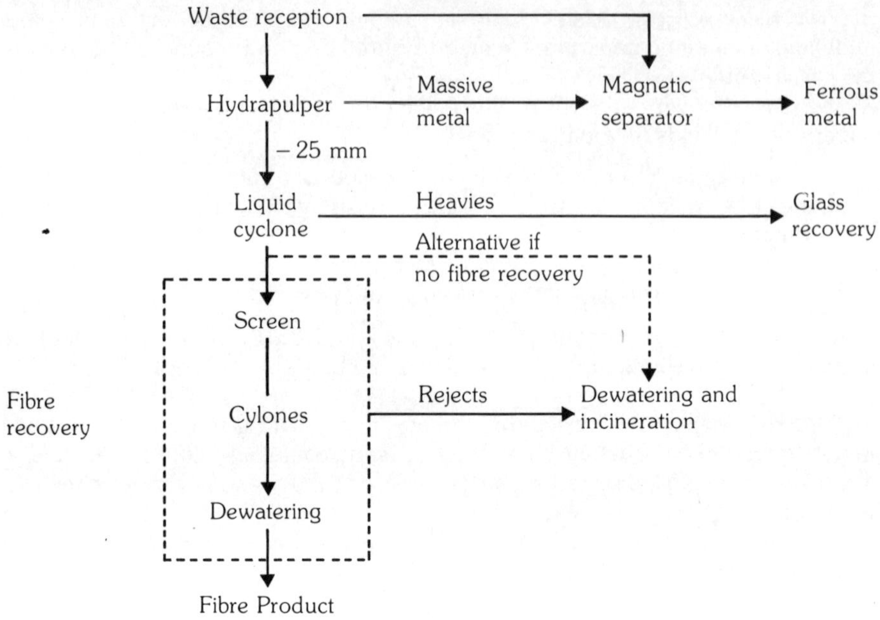

Fig. 9.16 Black Clawson "Hydrasposal" process.

a large tub with a high speed cutting rotor at the bottom; underneath the rotor, the pulp is discharged through openings with a diameter of 1.9 cm. Cans, large pieces of glass and stone are rejected through another aperture and collected by a 'junker', i.e. a continuous bucket elevator. The rejects are cleaned by recycled water introduced through the junker, washed in a rotary drum washer and sorted.

The impure pulp can gradually be purified to a paper fibre product by the following series of operations.

1. Large inorganic particles are separated by means of a low-pressure hydrocyclone. After elutriation of organic material, the reject stream forms a concentrate of glass and aluminium.
2. The paper is defibred in a VR classifier, consisting of a high-speed rotor surrounded by a heavy screen grate with 0.3 cm perforations. The classifier defibres the paper by a shearing rather than by a cutting action and eliminates non-fibrous materials, such as leather, plastics and twigs.
3. The pulp slurry is screened on a cylindrical pressure screen with openings of 0.15 cm. The openings are cleaned by the alternative pressure and suction, developed by a rotor turning inside the cylinder. The refined pulp passes into a conventional hydrocyclone, which eliminates inorganic dirt, shivers and slivers.
4. Finally the pulp is concentrated on an inclined screen, which retains only the longer fibres. The larger part of the water passes through the screen, together with shorter fibres, fillers, pigment and fine dirt.

The recovered pulp is fed into an inclined thickener, consisting of a perforated cylinder with an internal screw, which elevates the pulp and discharges it into a cone press.

Various subsystems have been tested for the recovery of aluminium, glass and nonferrous metals.

A first pilot plant was constructed in 1967 with a capacity of 0.5 tonne/h. A demonstration plant with a nominal capacity of 150 tonne/day has operated since 1971 at Franklin, Ohio, with support of EPA. A plant with similar capacity operates since 1975 near Tokyo. A 2000 tonne/day plant at Hempstead, NY is at present in commission.

The Black-Clawson system is the only completely wet refuse-sorting system currently in operation.

Other American Systems

The National Center for Resource Recovery (NCRR) is a non-profit-making association, which compiles and disseminates information on recycling in general and on refuse sorting operations in particular. Among the members of the association there are also companies interested in recycling systems and reclaimed products, e.g. equipment or packaging manufacturers, system developers and recycling enterprises.

The first sorting system operated by NCRR emphasized the production of a pelletized fuel. The equipment was installed around a pre-existing shredder for bulky wastes, situated at the site of a refuse incinerator in Washington, DC. The system made it possible to evaluate equipment for air classification, metal recovery and pelletizing of WDF (Fig. 9.17).

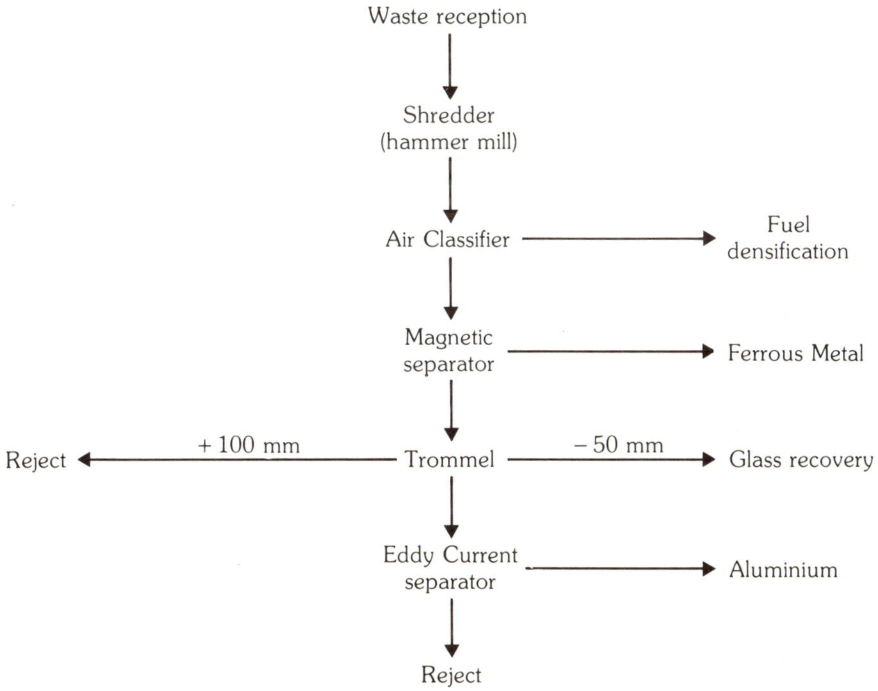

Fig. 9.17 NCRR-Washington refuse sorting system.

Since 1973, NCRR has been involved as a technical consultant in the design, construction and operation of a resource recovery plant. The (potential) output of the system is: WDF, ferrous metal, glass, aluminium, nonferrous metals.

The technical reports issued periodically by NCRR are probably the most reliable, detailed and unbiased information available on refuse sorting.

Occidental Research Corporation has developed a refuse sorting system, one of the end products of which was to be used as a feedstock for a proprietary flash pyrolysis system. Other recuperated products are ferrous metal, aluminium and glass.

A demonstration plant of a capacity of 200 tonne/day was constructed at El Cajon, Ca., but operated only for a limited number of test runs.

Conclusion

The selection of a particular sorting system depends on:

1. the nature and composition of the wastes to be processed,
2. the local market specifications,
3. plant capacity.

In a few years many more data will be available on the performance of separation equipment and systems, and on the sales potential of the fractions which concentrate paper, plastics, ferrous metal, organics and dense materials. Generally these fractions should be further refined and cleaned in order to satisfy current market specifications for secondary materials.

MARKETS FOR RECOVERED MATERIALS

Secondary raw materials are in a permanent competition with virgin materials. In normal conditions of supply and demand they survive by offering a fair and consistent quality, fixed by national or international specifications, at an attractive price.

The value of these secondary materials (see Table 9.2) varies with time, place, quantity, purity and form. Moreover, they are often of the same order of magnitude as the cost of transportation from a refuse sorting plant to the ultimate user.

The main product from refuse sorting in W. Europe will probably be a *mixed*

TABLE 9.2 Indicative Value of Some Commodities Available in Refuse

US	$/tonne
Paper	15 – 80
Plastics	not available
Compost	0 – 8
Fuel	10 – 20
Glass	10 – 25
Ferrous metal	12 – 80

wastepaper fraction. This product can be used in the manufacture of board, fluting medium for corrugated paper, and a number of building materials, such as asphalt-impregnated flat or corrugated roofing medium.

Newsprint could possibly be extracted separately. The mechanical properties of this paper are rather low, but it can be recycled in new newsprint after de-inking. It generally commands a better price than mixed wastepaper.

Common impurities of mixed wastepaper are plastic film and bottles, rope, textiles, wood, metal, glass and building materials. A number of special types of paper are undesirable, because they are coated, treated or laminated with wax, oil, grease, carbon, plastic film, aluminium foil, asphalt, varnishes, etc. Mixed wastepaper from central sorting is normally obtained by air classification. Plastic film can be largely eliminated after humidification of the paper (TNO), shrinking of the film during a heat treatment (US Forest Products, Fläkt) or selective shredding (Sorain, Enadimsa, Ebara). The removal is completed during wet pulping: the plastics accumulating in the pulper can be extracted manually, hydraulically and also mechanically, by installing a moving extraction line in the pulper. Small impurities are later eliminated from the pulp by means of hydrocyclones or related centrifugal separators. Non-pulpable materials are retained on screens during pulp purification.

The market for mixed wastepaper is relatively large, but fluctuating strongly. Stockpiling is not a practical means of stabilizing the market over long time periods, because wastepaper is bulky, flammable and degradable.

A potential much larger market for the light fraction, obtained by air classification, is as a *fuel* or a *fuel supplement*. The required fuel characteristics vary with the type of stoker and furnace used, but relatively homogeneous dimensions and a low moisture and ash content are always desirable. Moist combustibles are slow to ignite and reduce the capacity of the furnace. When the ash content is high the ash generating rate may exceed the existing ash handling capacity. A low melting point of the ash leads to boiler tube fouling and corrosion and to slagging of refractories.

Air-classified refuse can be compacted and transported to existing pulverized-coal-fired boilers or furnaces, such as cement kilns. Densified or pelletized material can be fired in small, mechanical grate boilers, but pelletizing costs a large fraction of the value of the fuel. Powdered waste-derived fuel, such as ECO-FUEL II, has a wide range of potential applications.

The *ferrous metal* extracted from refuse largely consists of can stock. It is contaminated with organic matter and paper. Moreover, the can is covered with tin and lacquer and contains a soldered seam and possibly an aluminium opening lid. This composite material can be detinned to yield a premium steelplate scrap and a small amount of tin, used in foundries, or converted into reinforcing bars.

Cullet can be re-used in the manufacture of glass containers. The main contraries are ceramics and stones, metals and organics. Colour-mixed cullet can be re-used in green and brown bottles; colour sorting somewhat augments its sales potential and its value.

CONCLUSIONS

In this survey of current separating technology and systems, several important aspects of refuse sorting have not been covered, such as:

1. The receiving, storage and internal conveying of refuse,

2. pollution abatement and safety considerations,
3. guidelines for construction and maintenance that take into account the unusual properties of refuse as a raw material.

In a few years, much more operating experience will be available from the various demonstration plants, with regard to the technical and economic feasibility of extracting secondary raw materials from refuse.

REFERENCES

1. *Resource Recovery and Conservation*, a Journal of Elsevier Scientific Publishing, Co.
2. *Conservation and Recycling*, a Journal of Pergamon Press.
3. *NCRR Bulletin, The Journal of Resource Recovery*.
4. Technical Reports by the NCRR, 1211 Connecticut Ave. NW, Washington, DC 20036.
5. COMMISSION OF THE EUROPEAN COMMUNITIES, DG XII — Research, Science, Education — Raw Materials R&D — Studies on Secondary Raw Materials I — Household Waste Sorting Systems — Brussels, Jan. 1979 (compiled by Warren Spring Laboratory — Construzionie Impianti SPA, Umweltbundesamt and Bureau de Recherches Géologiques et Minières).
6. THOMÉ-KOZMIENSKY, K.J., (Ed.) *Recycling*, Berlin 1979. FREITAG, E., Verlag für Umwelttechnik, Berlin.
7. JÄGER, B., and Thomé-Kozmiensky, K.J. (Ed.) *Materialrecycling aus Haushaltsabfall*.
8. ALTER, H. and Horowitz, E. (Ed.) *Resource Recovery and Utiilization*, ASTM, Special Technical Publ. 592, 1975.
9. *Proceedings of the Mineral Waste Utilization Symposium*, US Bureau of Mines — IIT Research Institute.
10. BUEKENS, A.G., *Resource Recovery and Conservation* **3** (1978) 275 — 306.

Alfons G. Buekens graduated as a Chemical Engineer (1964) and a Doctor in Applied Sciences (1967) at the State University of Ghent. His research work at this University was devoted to the study of the kinetics and mechanism of the thermal cracking of hydrocarbons. He was awarded the Main Award of the Royal Flemish Association of Engineers (1965) and the Robert De Keyser Award, donated by the Belgian Shell Co. (1966). Since 1969 he has been a professor at the Vrije Universiteit Brussel, where he teaches various courses related to Chemical Technology, Environmental Engineering and Industrial Satety. His research work is centred on problems of applied kinetics and waste disposal, generally involving thermal or catalytic conversion methods. He acts as a consultant for International and Governmental Agencies and private Industry, and is the author of one book on Solid Wastes and about 40 publications and numerous reports.

10 Gas Production from Landfill

O. Tabasaran

THE ORIGIN OF LANDFILL GAS

Municipal waste, as a first approximation, can be defined as a mixture of water with solid organic and mineral matter, including fats, carbohydrates and proteins. The water content varies, according to the time of year, between 25 and 50% by weight. If the refuse is properly deposited, a state is reached relatively quickly in which free oxygen and light are completely, or nearly completely, excluded: the metabolic process becomes anaerobic when the entrapped reserves of air are used up. The initial decomposition results mainly in acids, but also in alcohols, aldehydes and carbon dioxide, which are partly soluble in water. This basal metabolism is accompanied by constructive metabolism which synthesizes cellular substances. The decomposition products resulting from the first stage are attacked by specialised microorganisms, the methane bacteria which, in the presence of carbon dioxide, hydrogen, carbon monoxide and certain short-chain organic compounds, are able to produce methane.

If it is assumed that the three main factors in the process, the substrate, the microogranisms and the most important gaseous products, methane and carbon dioxide, can be quantified with their carbon content, an understanding of the metabolic process can be obtained by means of a decomposition model, as shown in Fig. 10.1, using 'organic carbon' as a parameter; the course of the gas production curve can be expressed, with sufficient accuracy, by

$$dC_g/(C_c - C_g) = kdt$$

where C is the carbon content converted to gas in time t, t is the reaction time, k is the reaction coefficient, and C_c is the total convertable carbon.

The anaerobic decomposition of organic compounds proceeds until the carbon accessible to bacteria, has been completely converted into sewer gas, microorganic material and other products associated with metabolic breakdown, provided that essential trace elements have not reached a minimum level. The gas development then reduces within the autolysis-regulated mass. As the autolysis is very slow, it can be regarded, in the context of this study, as negligible.

The methane bacteria and other acid-splitting organisms involved in the gas-production phase are obligatory anaerobes. They cannot live in the presence of free oxygen. They function best at a pH of 7.0, and are sensitive to higher concentrations of acid and to temperature variations of more than 2 °C. The water content of the substrate is also important; maximum activity occurs at moisture contents over 85%. The composition of the nutrient material and the extent and rate of decomposition also play an important part.

An improved substrate composition (expressed, for example, as the C – N ratio) gives favourable conditions for the development of anaerobic organisms[1].

Taking into account the climatic conditions in Central Europe and the normal composition of municipal waste, it can be stated in general that the moisture content does not normally present any obstacle to biochemical decomposition. Futhermore, the leach water acts as a transport medium for dissolved organic and inorganic substances and causes levelling of the composition of the substrate.

QUANTITY AND COMPOSITION OF LANDFILL GAS

Continuous conversion to gas of 1 kg of organically bound carbon, following the gas laws and under standard conditions, will give a gas volume of 1868 m^3, composed predominantly of methane and carbon dioxide. Under natural

Recovery of landfill gas.

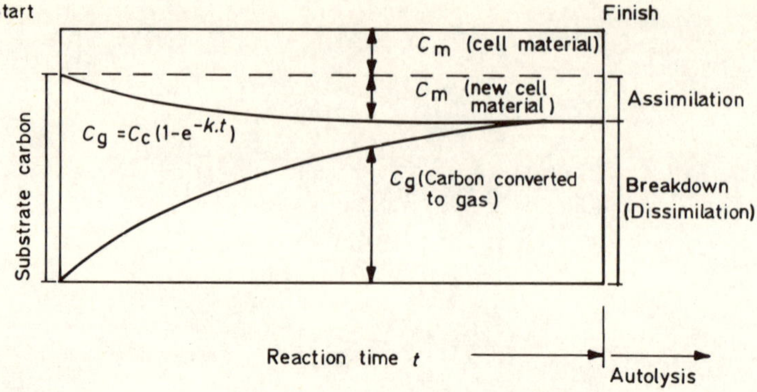

Fig. 10.1 Model for course of anaerobic breakdown with organic carbon as a parameter.

conditions, the specific quantity of gas that can be obtained is lower, since, depending mainly on the temperature conditions, only a certain proportion of the carbon content of the substrate can be converted; the remainder is used to produce new cell material. (See Fig. 10.1).

The relationship between gaseous and assimilated carbon follows an approximately linear course with rise in temperature in the mesophilic range, because the rate of assimilation is increasing here more slowly than the rate of gas formation. This relationship can be expressed, sufficiently closely, by the equation:

$$C_c = C_T (0.014\, T + 0.28)$$

where C_c is the total carbon convertible to gas over a long (infinite) period of time (neglecting autolysis), C_T is the total amount of carbon compounds in the substrate, and T is the temperature (°C).

Table 10.1 shows an example of the average proportions of the different types of material in the municipal waste of Stuttgart, Heidelberg, Karlsruhe, and the rural districts of Reutlingen and Tübingen. In these, biochemically degradable materials predominate, including kitchen and food wastes, paper, leather, cardboard and wood. Kitchen waste can be regarded as easily decomposable. The proportion of kitchen waste in Germany varies between 29 and 52% (by weight) of the total, whilst paper, which has an intermediate rate of breakdown, forms 13 - 23%. The most difficult materials to decompose, leather, cardboard and wood, form 4 - 10% of the total. In waste of average composition, the organically bound carbon content C_T is about 200 kg/tonne. The gas developed in the waste over a long period of time (G_c, the maximum yield) can be calculated, using Equation 10.2, as

$$\begin{aligned} G_c &= 1.868 \times C_c \\ &= 1.868 \times 200\, (0.014\, T + 0.28) \\ &= 5.23 \times T + 104.61 \quad \text{m}^3/\text{tonne of waste} \end{aligned}$$

The temperature in the waste, shortly after deposition, reaches a high value as the metabolism of the aerobic microorganisms produces free energy in the form of heat. It then declines. The average level reached over a number of years is household waste, landfills, and depending on climatic conditions and deposition technique, etc., is between 7 and 20 °C. T = 15 °C can be taken as an

TABLE 10.1 Examples of Composition of Municipal Refuse

Material	Stuttgart 1974	Reutlingen district 1974/1975	Tübingen district 1974/1975	Heidelberg FJ 1976	Karlsruhe H 1978
Metals	5.3	4.4	4.7	5.2	3.4
Glass	9.9	13.4	13.8	13.8	11.7
Inert material	1.2			2.0	0.5
Leather, cardboard, wood, etc.	4.1	5.6	5.2	10.4	5.5
Textiles and plastics	6.2	8.2	7.6	8.4	10.3
Dust, ashes, etc.	6.2	9.5	10.7	7.2	7.0
Kitchen waste, food residues, etc.	52.4	45.8	44.3	29.6	43.0
Paper	14.7	13.1	13.7	23.4	18.4
Total	100.0	100.0	100.0	100.0	100.0

approximate average value, although individual readings of waste temperature of over 30 °C have been recorded. Thus,

$$G_C = 5.23 \times 15 + 104.61$$
$$= 183 \quad m^3/\text{tonne of waste} \qquad (10.3)$$

Experience to date justifies the comclusion that the production of methane starts intensively about a year or two after deposition and continues for about 20 years. Gas can be detected after the end of the 20 year period, often up to 50 or 75 years. For purposes of prognosis and comparison, however, it can be taken that over a period of 20 years a stable state exists, during which time about three quarters of the final volume of gas has been formed, and Equation 10.1 can be regarded as valid. Replacement in the equation gives:

$$G_t = 183\,(1 - 10^{-k.t}) \quad m^3/\text{tonne waste}$$

where G_t is the quanitity of gas produced in t years; for example,

$$G_t = 183\,(1 - 10^{-0.03t}) \quad m^3/\text{tonne waste} \qquad (10.4)$$

where t is the time in years. Equation 10.4 is expressed graphically in Fig. 10.2

The curve for specific annual production of gas is steeper at the beginning than later on. The equation for this (see also Fig. 10.3) is

$$G_t = (13.2)\,(10^{-0.03t}) \quad m^3/\text{tonne/year} \qquad (10.5)$$

This calculation shows that a tonne of waste material gives off about 12.3 m^3 of gas during the first year, about 11.5 m^3 in the second year, about 10.7 m^3 in the third year and, for example, 6.6 m^3 in the tenth year.

In controlled deposition, the highest rate of gas emission theoretically occurs at the termination of the landfilling. It then decreases gradually. The curve for

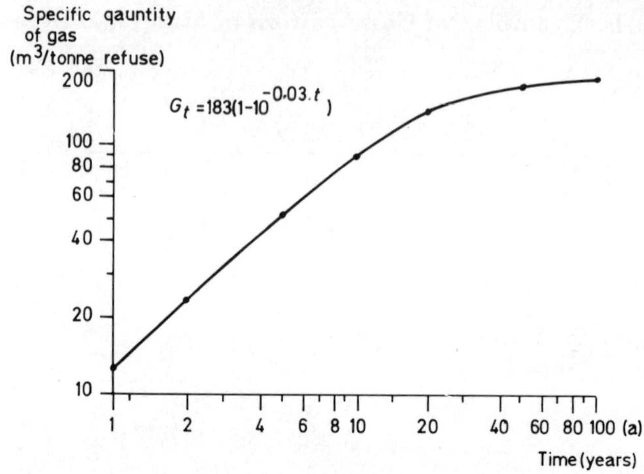

Fig. 10.2 Total curve of specific quantity of gas.

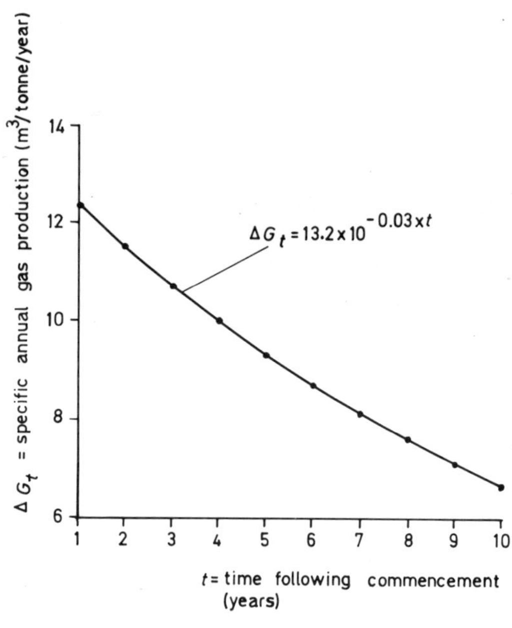

Fig. 10.3 Curve of specific annual gas production

gas production per tonne of household waste in relation to time, with a landfilling duration of 10 years, is shown in Fig. 10.4.

The landfill gas produced under strictly anaerobic conditions contains approximately 54% methane and 45% carbon dioxide, together with smaller quantities of hydrogen, hydrogen sulphide, carbon monoxide, mercaptans, water vapour, particulate matter, etc. Typical composition is shown in Table 10.2. The presence of nitrogen, oxygen and inert gases indicate the presence of air residues. If, after closed and compacted deposition, concentration ratios for

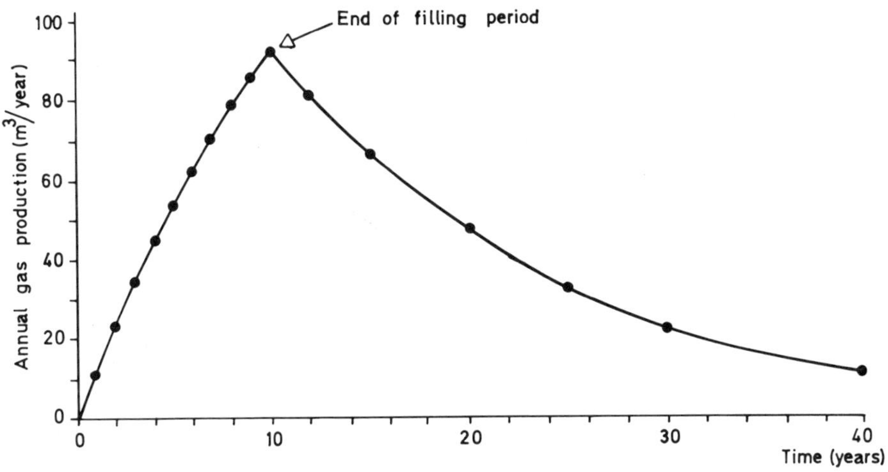

Fig. 10.4 Theoretical gas production over a period of 40 years. Annual quantity of waste = 1 tonne, filling time = 10 years, total quantity of waste = 10 tonne/year.

CH_4 to CO_2 of 0.9 to 1.2 are obtained, the cause of the raising (or lowering) of the CO_2 or CH_4 content is an air breakthrough.

Knowledge of the actual composition of the landfill gas allows conclusions to be reached on the type of biological process from which it is derived. During the first two weeks following deposition, relatively high amounts of nitrogen and oxygen are recorded. The CO_2 concentration reaches its maximum value in just about two months and thereafter stops at approximately 45% (by vol.) for about two years. The methane production, after a delay of two months, rises over the course of two years to a level over 50% 'ɔy vol.). The presence of hydrogen is mostly not noticeable during the first year.

TABLE 10.2 Example of Composition of Typical Landfill Gas[3]

Gas		Vol. %
Methane		53.283
Carbon dioxide		45.588
Heptane		0.292
Nitrogen		0.272
Octane		0.272
Oxygen		0.070
Nonane		0.064
Hexane		0.128
Hydrogen		0.056
n-Pentane		0.014
iso-Pentane		0.010
Propane		0.007
n-butane		0.006
iso-Butane		0.004
Hydrogen sulphide		0.002
Calorific value	5.162 kcal/m^3	21.612 kJ/m^3
Density		1.017 kg/m^3

The course of the decomposition in the deposited waste results in the formation of complexes. The composition of the refuse, temperature, access of air, moisture content, pH, alkalinity, formation of toxic compounds, etc., in the immediate surroundings of the microorganisms control the rate of formation of gas. Other factors, such as air temperature, atmospheric pressure, type of deposition and nature of surface layer, amount of atmospheric precipitation, topography, hydrology, etc., affect the rate of infiltration. The effect of admixture of air is particularly noticeable in the upper 20 - 25 cm of the mass. For example, at the Lemberg landfill site in the Ludwigsburg district, gas taken by means of probes pushed in at various depths and analysed[4] gave the results shown in Table 10.3.

TABLE 10.3 Gas Analysis at Different Depths

Depth (cm)	Gas components (% vol.)			
	CO_2	O_2	CH_4	N_2
20	5.1	17.4	3.5	73.8
25	9.7	16.1	8.8	65.4
30	32.8	4.2	35.6	27.4
45	37.7	2.6	41.1	18.6

LANDFILL GAS AS A DANGER SOURCE

Landfill gas can, in some cases, constitute a potential danger in the form of detonations, fire, odours, contamination of ground water and, not least, damage to vegetation. A porous type of final covering allows the gas to escape relatively evenly, so that no significant concentration is formed at the surface. On the other hand, covering with dense material such as clay prevents uninterrupted loss of the gas, which by diffusion and convection collects under the surface layer; it then takes the easiest escape route through the more porous areas and through holes and cracks, the makes its appearance sometimes in the most undersirable places. A gas-carrying fissure in a particular place may not be permanent; it may come and go as a result of climatic effects.

A number of regrettable incidents are known [5-12] where landfill gas has seeped into a closed room and exploded, causing damage and injury.

Recently, more attention has been directed to problems resulting from the effect of landfill gas on plant growth. Methane is certainly not toxic; however, with its small molecular size it can expel oxygen and cause suffocation of the roots of plants. Most cultivated plants are not capable of growing undisturbed with less than about 5% of oxygen in the region of the roots. Also, at higher temperatures in the material and the upper surface, the presence of large concentrations of hydrogen sulphide or carbon dioxide can injure plant life[13-15].

It should be mentioned that landfill gas in the presence of calcium bicarbonate can raise the calcium hardness, and the related pH value, of water, as well as initiating corrosive properties as a result of the dissolved carbon dioxide.

THE TEST-LANDFILL 'AM LEMBERG' AND GAS COLLECTION FACILITIES*

Four years ago the University of Stuttgart started a research programme with the large landfill 'am Lemberg' near Stuttgart, to determine basic information on the origin, movement and recovery of the gas. Today, many are interested in the use of landfill gas; however, this was not the case at the beginning of the study. Times have changed.

The landfill 'Am Lemberg' has two parts. Part I is nearly filled and covered and has a content of about 1.6 mio m^3 of waste; part II began in 1977.

The landfill has an impervious bottom with a drainage system. it is built up to 2 m lifts with the help of special compactor machines. At first, the layers were covered daily, mainly by foundry sand, but this was stopped because the sand caused leachate accumulation.

The landfill is filled mostly with rubbish and waste of small industry, but also with sludge cake from sewage treatment plants. The waste is discharged between circular berms 3 - 4 m thick and consist of clay and soil. Finally, the landfill was covered with 2 m of soil and landscaped. Since the beginning of the second part of the landfill, gas standpipes were built because great damage caused by gas on the first part of the landfill was noticed. These vertical pipes consist of perforated concrete rings with an inside diameter of 1 m with a concrete casing, 0.3 m in diameter in the middle, backfilled with one-inch rocks. The distance between the gas wells is 100 m. In part I of the landfill, gas collection systems are being installed. To test different methods, a gas collection network was installed below the covering soil, and gas wells for withdrawing gas were drilled. As a result of the tests, the study team now favours gas wells with a diameter of 1.3 m, reaching down to the bottom of the landfill, because gas pressures of up to 30 cm water were found.

For research purposes, several testing facilities were also installed in the landfill 'Am Lemberg' mainly using flexible tubes. Temperature, pressure and gas concentration were measured, and tracer tests carried out regularly both there and at the wells. Several probes at different depths were also worked on to determine changes in gas concentration, gas horizon and flow characteristics. At part I of the landfill, probes were installed for permanent control. Experiments like blowing air into the landfill, extracting gas and using tracers produced interesting results. Furthermore, permanent pumping tests are made with an efficiency up to 240 m_3/h at gas wells of depth 10 m and diameter 160 mm. Pressure and gas concentration at the gas wells and surrounding pipes were measured red.

Plastic boxes were also installed on the surface of the landfill to catch the gas seeping upward, and it quantity was calculated. The measurements of gas concentration were carried out using a gas chromatograph built into a truck with an independent energy supply.

To summarise the results:

(i) At a new landfill, the development of gas can start within few months.

(ii) A gas concentration gradient develops on the surface of an uncovered landfill so that in this upper area methane can be found only in traces; this is the atmosphere-influenced zone. Here methane readings are low and eminating gas is nearly odorless. So some prefer the landfill technique that does not cover the landfill refuse and finally grows plants on top.

---

*This section was prepared by Dipl.-Ing. G. Rettenberger, University of Stuttgart.

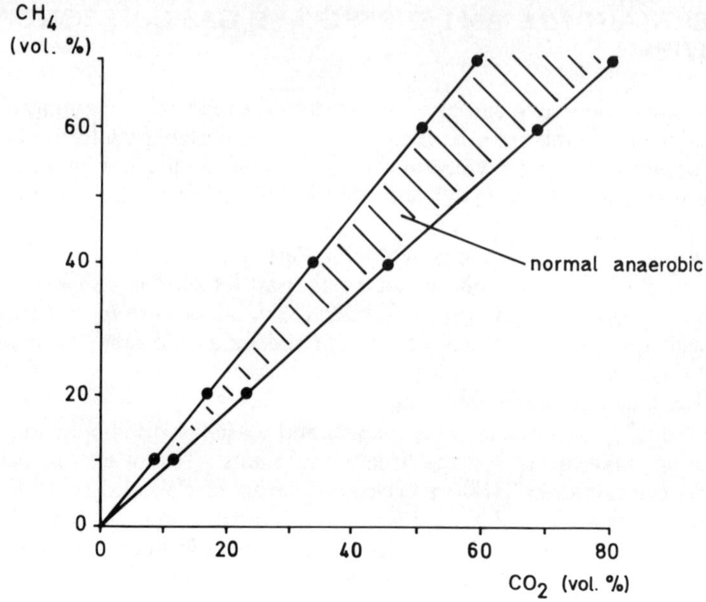

Fig. 10.5 Stable anaerobic range of proportions of CH_4 to CO_2 in landfill gas.

(iii) Gas accumulates below in covered landfills, and the pressure varies from being low at the slopes, to 10 - 20 times higher in the centre of the landfill.
(iv) The gas composition varied only a little (+1 - 2%) in 1.5 years: it was independent of the meteorological situation and we usually found 56 -58% methane, and 42 - 44% carbon dioxide. The gas was nearly saturated with water, the condensate was rather acidic (pH 4.3), and it also contained salts and organic solids.
(v) The temperature in part II of the landfill ranges from 33 to 40 °C, whereas in part I we measured up to 50 °C.
(vi) Gas could be found in all plastic boxes in spite of a 3 - 4 covering of soil. Maximum values reached $61/m^2$ h.
(vii) The radius of influence of the gas vacuum was about 50 m with a downhole vaccuum of 60 cm water column.
(viii) The gas flow in the unisotropic waste is nonhomogeneous.
(ix) Gas resources in different landfills can vary a lot. This means that in each case dynamic tests must be carried out.

At the test landfill there are now five gas wells. For their construction the following criteria were considered: the gas wells

are 100 m apart,
reach down to the bottom of the landfill,
have large diameters,
can be dewatered,
have a perforated part that ends 4 m below the surface,
are made of corrosion-resistant material.

From the test data, a daily gas production of 0.06 - 0.15 m^3/m^3 volume of waste

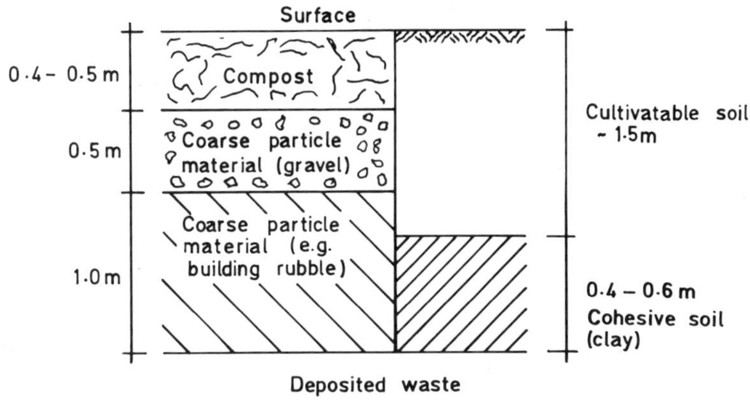

Fig. 10.6 Cross-section through large-surface filter.

was calculated. Even if maximum capture of 20% of the gas production is assumed, 720 m^3/h is still obtained from the landfill. This appears to be enough for practical utilization. Plans for profitable utilization of the landfill gas are being worked on. The best way seems to be either to utilize the gas to supply nearby greenhouses or to produce electricity. At first it should be possible to deliver 300 - 400 m^3/h for the coming 18 years.

In Europe there is practical experience of using landfill gas only in boilers, although making electricity with sewage treatment gas is common. In Switzerland a special boiler has been developed in comibination with landfill gas recovery facilities to heat the landfill support buildings. The advantages of this boiler is that it is able to use gas that contains only 22% methane and the flare burns even with 18% methane. Other techniques have not yet been used.

DEODORIZING OF LANDFILL GAS

The gaseous end-product of the anaerobic biochemical processes contains, in addition to methane and carbon dioxide, a number of intensively odorous substances. The existence of cavities leading up to the surface of the landfill presents a risk of breakthrough of odours, which can, however, be removed by thermal or biochemical oxidation. In thermal oxidation the destruction of the gas at a temperature of at least 800 °C, with a dwell-time of not less than 0.3 s, is usually specified. Biochemical oxidation can be carried out by passing the gases through a biofilter. The gas is absorbed on to the surface of the particles of a biologically active filter material, and is finally dispersed microbiologically. The filter surface must be regenerated in a continuous process in which a balance is achieved between adsorption and respiratory oxidation.

The presence of moisture and elementary oxygen is a basic requirement for the aerobic metabolic conversion. Experiments show that the biofilter can achieve deodorizing of landfill gas when there is an average water content in the filter material (mature compost) of 50%, and a gas-feed mixture of 1 part (by volume) of air and 10 parts (by volume) of landfill gas, and a loading of 5 m^3/m^2 is not exceeded[4].

SPREADING OF LANDFILL GAS

The gas produced by the microbical breakdown of the components of refuse spreads itself out in various directions, mainly by diffusion and pressure flow.

Diffusion flow results from the movement of gas from zones of higher concentration to zones of lower concentration; this migration is caused by the partial-pressure gradient.

Pressure flow occurs when gas moves from a zone of higher pressure to one of lower pressure.

The entry of air into the landfill material follows the same laws of diffusion and mass transport. Through faults such as porosity of the covering layer, the pressure built up in the body of the landfill comes into balance with atmospheric pressure. A discharge of gas can be initiated by turbulence forces such as wind.

Diffusion ceases when a concentration difference between the atmosphere and the gas in the landfill no longer exists, or when the pressure levels out. In covered landfills, the gas flow is in general obstructed by a pressure of a few millibars below the surface layer. If the gas is removed, that is if a reduced pressure is created, by the use of an airpump, fan or blower, the balance is disturbed and gas flow is set in motion.

Mathematical models for the spread of landfill gas can be constructed along the above lines, taking into consideration the usual basic factors applicable in gas transport studies in soils.

TECHNICAL MEASURES FOR CONTROL OF LANDFILL GAS

In the control of landfill gas, use is made of the fact that the gas moves mainly in the direction of minimum resistance, and more quickly into porous material. A distinction is made between measures taken on the landfill and those taken in the neighbourhood of the landfill. In completed landfills, satisfactory results can be obtained by driving perforated pipes into the mass. Application of vacuum in the pipes increases the collecting range and the volume of gas extracted.

When new landfill is being deposited, the arrangements for gas control can be laid in from the start in the form of vertical shafts or horizontal channels.

Gravel-filled pits can be used for the protection of particular objects in the vicinity of landfills, and can, if required, be fitted with pipes connected to pumps. There is also the possibility of placing perforated pipes between the landfill and the object to be protected. Pits filled with non-porous material such as clay can be used as barriers; even hybrid systems using both barriers and gravel-filled collector pits are possible.

MANAGEMENT PROCEDURES FOR CONTROL OF GAS PRODUCTION

The main procedures available for acceleration of the biochemical process in a landfill are homogenization of the deposited wastes, addition of sewage sludge and attempts to control the water content within the material. Homogenization can be carried only by shredding the refuse before deposition. However, the use of compactors for tipping also has the effect of homogenizing to some degree. The simultaneous deposition of sewage sludge and refuse gives an increased initial water content in the mixture and increases the supply of nutrients to the microorganisms, resulting in more intensive gas generation. Mixing the refuse and sewage sludge together in a drum-type mixer, as in the composting process, also offers the benefit of a more rapid metabolic process.

If intermediate covering is used, or if the fill is built up section by section on a small area, the penetration of surface water is reduced and this results in a fall in the moisture content of the deposited material; a side-effect is a slowing down of the microbical metabolism. Conversely, the avoidance of thick intermediate layers and depositing on larger area sections promotes gas production.

An increased rate of gas production means an acceleration of the exothermic process by which energy in the form of heat is released, increasing the temperature of the surrounding material. This increase in temperature causes an increase in the rate of metabolic activity, so breakdown proceeds faster than assimilation, which again leads to an increase in rate and volume of gas production.

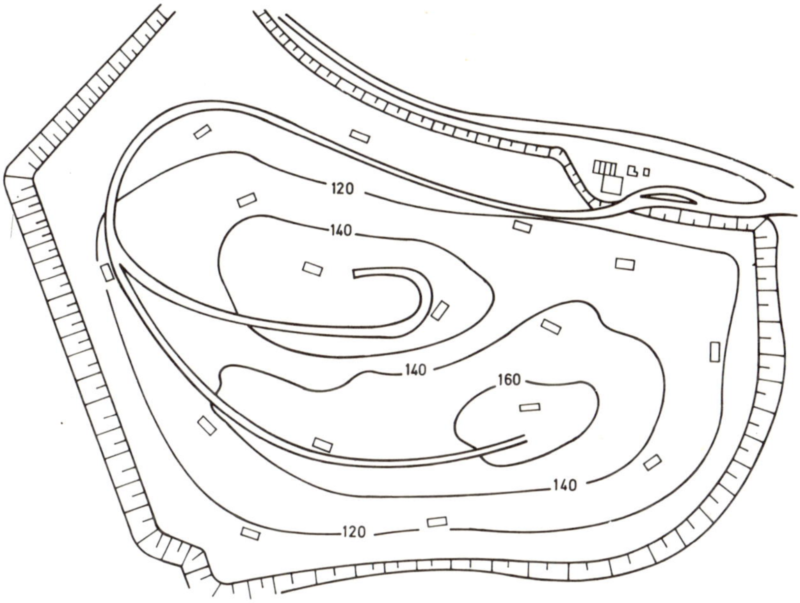

Fig. 10.7 Diagram to show location of large-surface filters on landfill.

The tipping or deposition technique used influences the extent and the speed of gas production, within certain limits.

SUGGESTIONS FOR DEGASSING OF LANDFILL BY MEANS OF LARGE-SURFACE-FILTERS

Previous experience shows that all forms of induced degassing bring problems, which often remain many years after the closing of the landfill.

Burning flares, even when there are no other risks left, cause the prolongation of safety precautions. In addition, any equipment that has been installed, such as suction machinery, requires maintenance and attention over a long period.

The expenditure can only, as a rule, be justified if the gas so obtained is utilized in some way.

Recent investigations indicate that, with porous surface layers, air penetrates the upper layers, either by diffusion or by the pressure differences caused by wind, thereby providing oxygen. With time, the amount of 0.5 m^3/m^2 h, which is sufficient to supply the oxygen required for biochemical deodorizing of 5 m^3/m^2 h of landfill gas, can be taken in. Accordingly, in normal cases a biologically active surface of 100 m^2 per hectare of landfill furface is sufficient to give a marked reduction in odour emission. Therefore, separate filters, evenly distributed, should be provided, covered not with ordinary covering material but with coarse inert material, such as gravel or broken stone, and with a top layer about 0.4 m thick containing composted refuse.

The covering layer on the areas between the filters should consist of about 1.5 m of cultivatable soil on top of about 0.5 m of cohesive material, e.g. clay. In the suggestion as sketched out in Figs. 10.7 and 10.8, the provision of a relatively

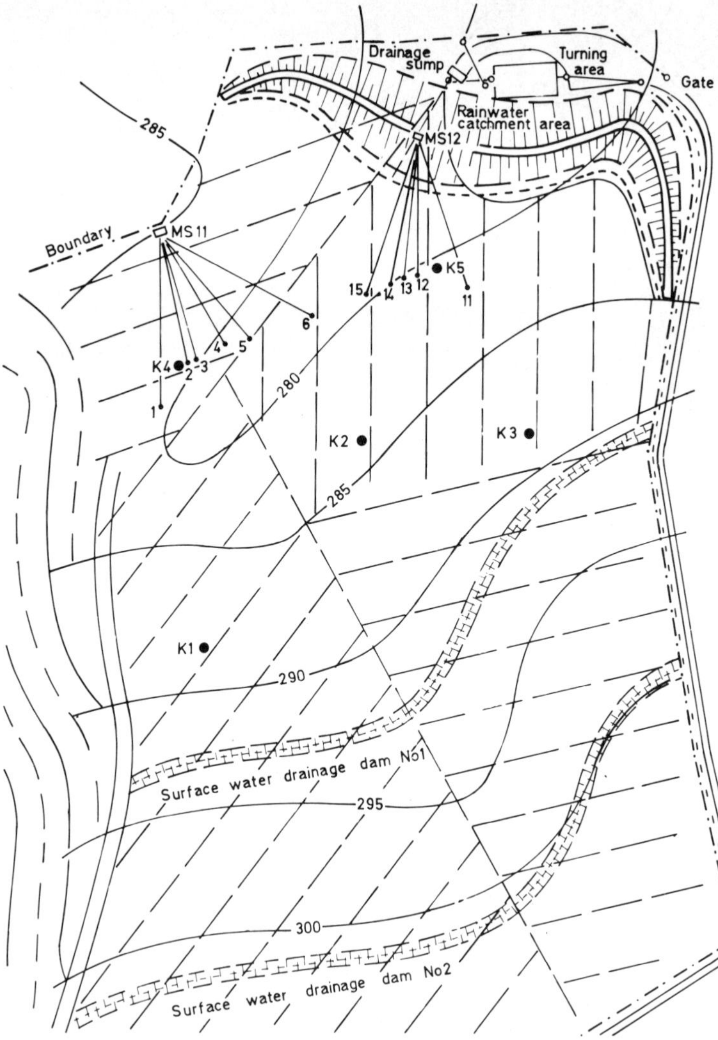

Fig. 10.8 Landfill site 'am Lemberg' part II.

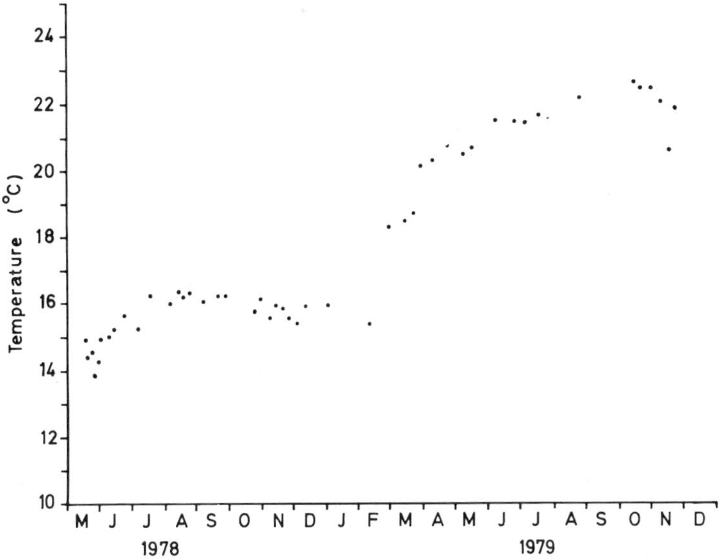

Fig. 10.9 Temperature of the Leachate (landfill 'am Lemberg' part III).

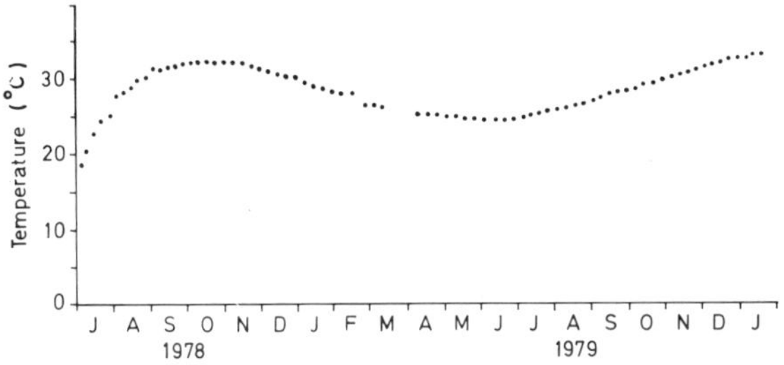

Fig. 10.10 Temperature of the deposited waste in landfill 'am Lemberg' (MSt 2).

thin, optimally compacted sealing layer is particularly important. The location of the large-surface filters on the landfill areas would depend on the shape of the area and be decided in consultation with specialists in re-cultivation, together with the question of the texture of the cultivatable soil. To ensure continuous operation, care must be taken to see that the filters do not dry out and that no cracks appear. Occasional surface redistribution and wetting may be necessary.

REFERENCES

1. O. TABASARAN, 'Uber Maßnahem zur Beschleunigung der Schlammfaulung', *Stuttgarter Berichte zur Siedlungswasserwirtschaft*, 33 (1977) Oldenbourg Verlag, München.
2. CDM, Inc. Pasadena, Calif., *A Case Study of the Los Angeles County Palos Verdes Landfill Gas Development Project*, July 1977, US Dept. of Commerce, PB 272241.

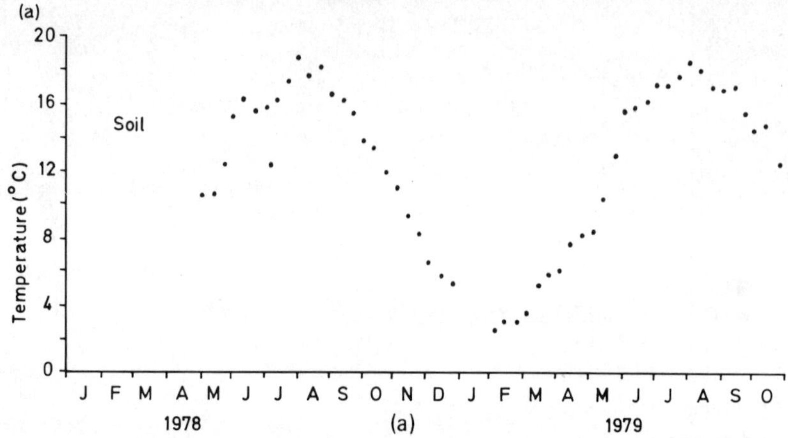

Fig. 10.11 (a) Earth temperatures and

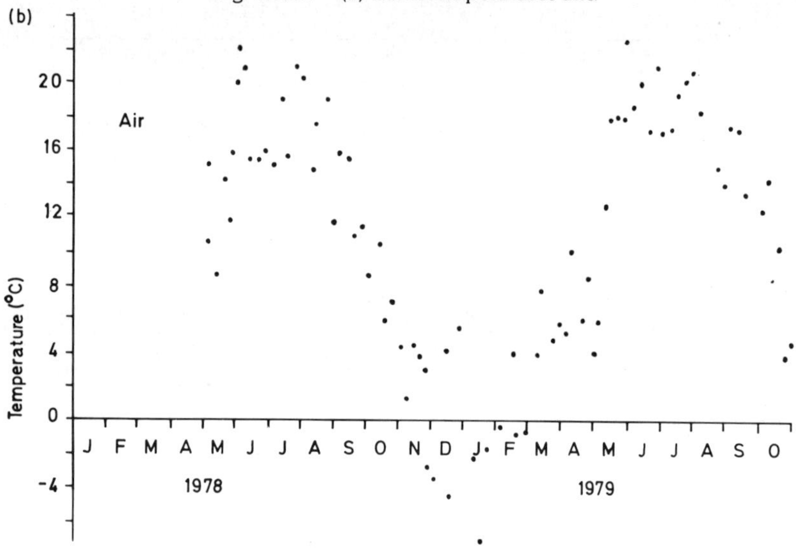

(b) air temperatures.

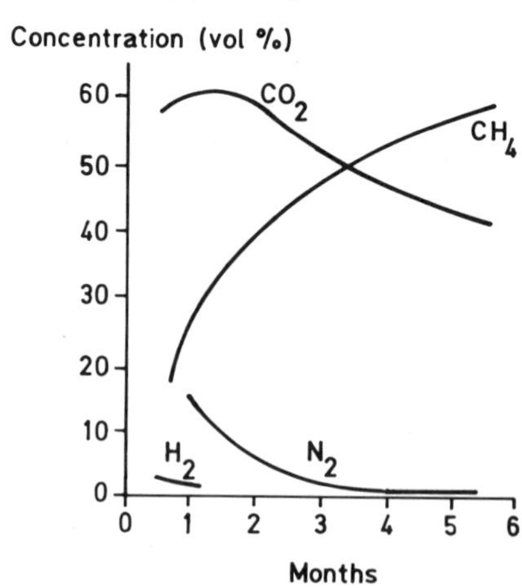

Fig. 10.12 The composition of the gas during the first six months.

3. FARGUHAR, G.J. and ROVERS, F.A., 'Gas Production During Refuse Decomposition Water; *Air and Soil Pollution*, **2** (1973), 483 - 495, Reidel, Dordrecht, Holland.
4. TABASARAN, O., RETTENBERGER, G. and AFFOYON, L., Einsatz von Biofiltern zur Deponiegasdesodorierung, *Müll and Abfall*, **5** (1979), Schmidt Verlag, Berlin.
5. HAGERTY, J., 'Gewinnung, Behandlung und Verwertung von Deponiegas, Beispiele aus USA', *Stuttgarter Berichte zur Abfallwirtschaft*, **7**, Schmidt-Verlag, Berlin.
6. ELIASSEN, R., 'Why you should avoid housing contruction on refuse landfills', *Engineering, New-Record*, **138** (18) (1947) 90.
7. MAC FARLANE, I.C., 'Gas Explosion Hazards in Sanitary Landfills', *Public Works*, May 1970, p.76.
8. PATRICK, P.K., 'Gestaltung and Betrieb von geordneten Deponien im Vereinigten Königreich', *Stuttgarter Berichte zur Abfallwirtschaft* **7**, Schmidt-Verlag, Berlin.
9. DENNIG, Rechtsdirektor, Main-Taunus-Kreis, persönliche Mitteilung.
10. SOWERS, G.F., 'Foundation Problems in Sanitary Landfills', *Journal of the Sanitary Engineering Division of ASCE* **94** (SA 1), Feb. 1968.
11. ZABETAKIS, 'Biological Formation of Flammable Atmospheres', *Report R.J. 6127*, US Dept. of the Interior, Bureau of Mines, Washington DC, 1962.
12. WINSTON-SALEM Dept. of Public Works. *An Evaluation of Landfill Gas Migration and a Prototype Gas Migration Barrier*, US Dept. of Commerce, Springfield, Va. 22161, USA.
13. PIERAU, H., 'Entgasung von Abfalldeponien', *Kommunalwirtschaft*, **4** (1975).
14. FLOWER, F.B. and MILLER, L.A., *Report on Investigation of Vegetation Kills Adjacent to Landfill New Brunswick, NY*, Cooperative Extention Serv., Rutgens University, 1969.
15. ELIASSEN, R., O'HARA, F.N. and MONAHAN, E.C., 'Sanitary Landfill Gas Control' *The American City*, Dec. 1957, p. 115.
16. TABASARAN, O., 'Uberlegungen zum Problem Deponiegas', *Müll und Abfall* **7** (1976) 204 - 210, Schmidt-Verlag, Berlin.
17. Tabasaran, O. and RETTENBERGER, G., 'Deponiegas-Untersuchungs ergebnisse', *Abfallwirtschaft* **1** (1978) 137 - 174, Eigenverlag Institut, für Gewässeregulierung, Landwirtschaftlichen Wasserbau und Abfallwirtschaft der TU Wien.
18. AHTING, D., Beispiele für Entgasungsmaßnahmen auf Deponien, *Stuttgarter Berichte zur Abfallwirtschaft* **9** (1978) Schmidt-Verlag, Berlin.
19. SCHLAITZER, J., 'Entgasung der Deponie Poppenweiler' *Stuttgarter Berichte zur Abfallwirtschaft* **9** (1978) Schmidt-Verlag, Berlin.
20. RYSER, W., *Gase aus geordneten Deponien*. Internationale Studientagung der VGL über Sickerwasser und Gase aus geordneten Deponien, Nov. 1978, Krattigen, Switzerland.
21. Umweltbundesamt, UBA-Materialien 7/78, *Gas in Abfallablagerungen*, Schmidt-Verlag, Berlin, 1973.

Professor Tabasaran is Head of the Department for Solid Waste Techniques and Chair-Professor for Solid Waste Management in the University of Stuttgart, Faculty of Civil Engineering.

11 Incineration of Refuse with Energy Reclamation

B. Enhörning

INTRODUCTION

There are five main ways of treating household waste (see Table 11.1). Of these five items, only the first three are relevant to the subject of this chapter. The direct combustion process is by far the most common and the closest example of using this system is the Volund incinerator in Palma de Mallorca where the well known Danish firm A/S Volund, Glostrup has recently started up an 8.25 tonne/h unit with full heat recovery. There are about 280 facilities operating on this basis in Europe, Japan and the USA and therefore this can be called a well established technique, rather than a technique in development, with the associated problems. Compared with experience from refuse-derived fuel (RDF) production and burning and pyrolysis techniques, incineration can be carried out without too much difficulty; the ultimate goal to destroy waste or reduce its volume and extend the life of landfill can be reached without too much uncertainty with respect to cost and availability.

TABLE 11.1 Methods of Treating Municipal Wastes

Process	Principal fuel or converted energy products
1. Direct combustion processes (mass burning or incineration)	Steam; hot or chilled water
Refractory furnace	
Waterwall combustion boiler	
Small-scale package incinerator	
2. Mechanical separation of solid combustibles (RDF)	
Dry process (shredding and air classification)	'Fluff' RDF Dust RDF Densified RDF
Wet process (hydrapulping)	Wet RDF
3. Pyrolysis	Low Btu gas Medium Btu gas Liquid fuel
4. Bioconversion	
Landfill	Methane
Anaerobic digestion	Methane
Acid hydrolysis	Methane, ethanol
Enzymatic hydrolysis	Methane, ethanol
5. Brayton cycle	Electricity/steam

INCINERATION

Fig. 11.1 shows a conventional incinerator with the possibility of full heat recovery. This particular unit raises steam to produce electrical power in a back pressure turbine and heat for district heating and is therefore run at comparatively high steam rates.

The most serious problem in running an incineration plant is disturbance of the burning process and blockage at both ends when the user is forced or fooled into receiving unsuitable waste, such as too much PVC, rubber, glass and chemicals, or bulky material, such as car scrap, barrels, bathtubs and refrigerators. This can only be avoided by teaching the deliverers or producers of municipal solid waste (MSW) certain rules for sorting out unwanted materials. The presorting of incoming waste seems to be necessary to avoid stoppages. Furthermore, all plants have to be furnished with some sort of bulky waste disintegrator, i.e. a mill or shredder. The crane pilot must be constantly aware of the risks of feeding unfit matter into the hopper but he must also have means to reduce its size within the limits of the actual measures. The common 'sneaking in' of risky wastes (explosives, solutions, cats and dogs) seems to be unavoidable but must, with the help of information and discipline, be kept to a minimum.

The most frequent disturbances in incineration plants seem to be the following:

(i) *Bunker fires*. These can quite easily be taken care of by sprinklers and water cannons. Sometimes it will be necessary to call out the fire brigade to use a foam extinguisher.

(ii) *Breakage* of the cables or wires connected to the *grab*, requiring replacement about 4-6 times/year.

(iii) *Damage to the moving grates*. The cause of the damage can be mechanical (heavy objects, spark plugs, nails), thermal (high-temperature spots caused by overload in the fire room) or clogging of the air cooling slots between the bars by melting metals (lead, fire).

(iv) *Corrosion* and *erosion* damage to the boiler tubes, especially to the superheaters.

(v) *Slagging* of *brickwork* in the fire room causes cracking and sometimes complete breakdown of the walls.

(vi) *Corrosion* and *cracking* of the electrodes in the electrical precipitators.

The value of MSW as a substitute for oil and coal has long been recognized. The heating value of waste has increased quite rapidly in the sixties and has now stabilized on a value of 2.5 – 3.0 MWh/tonne. This, together with the increased amount of plastics in the waste, has caused a temperature rise in the fire box, which in turn has made the design of the incinerator more like that of a common waterwalled coal-fired boiler. Also, the tubing in the nearest area above the moving grate had to be protected against wear and corrosion (reducing atmosphere) with the help of some sort of high-temperature-resistant padding. Fig. 11.2 gives an example of the type of protection for the most exposed tubes.

An alternative design can be the completion of the protection liner in the most exposed zone with a string of perforated, air-cooled high-quality cast iron sheets. A lot of incinerators have been modified in the last five years in accordance with this Swiss patent, with very good results. The main purpose of this design is to keep the reducing atmosphere away from the sidewalls and at the same time add as much oxygen as is needed to oxidize the volatiles fully before the gases come into contact with the fire box walls.

The corrosion of the steel surfaces, whether they are placed in the high-

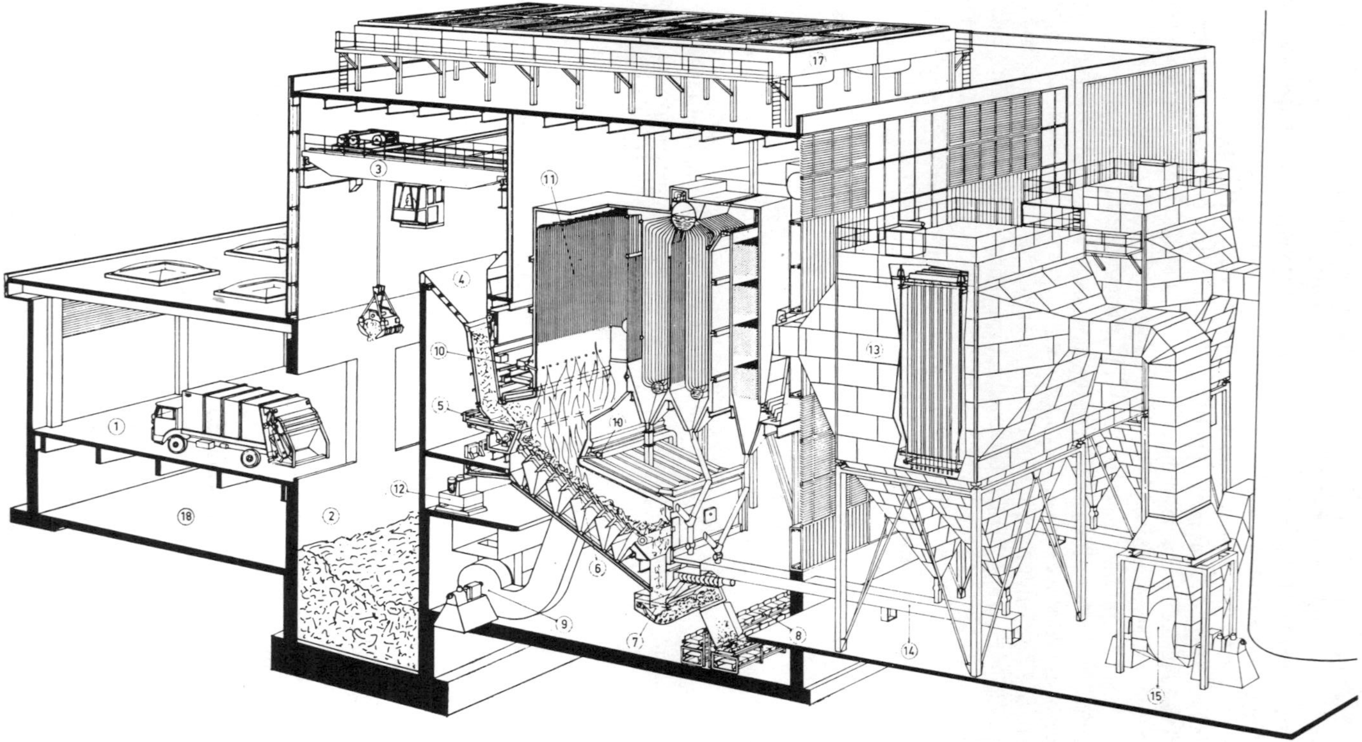

Fig. 11.1 The fourth incineration unit in the city of Nürnberg, manufactured by Martin Feuerungsbau, München.

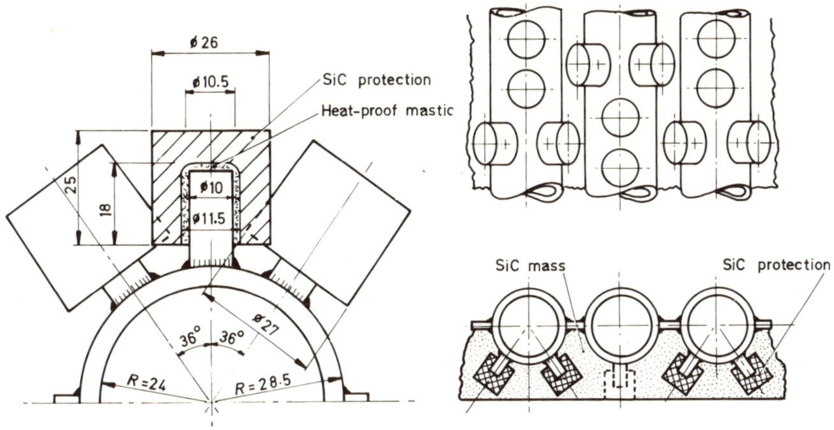

Fig. 11.2 Lining of boiler tubes with SiC (silicone carbide) protection caps and SiC padding. (Reproduced, with permission, from H. Stein, G. Gelsdorf and M. Schwalb, *VGB Kraftswerkstechnick* **59**, 4 (1979).)

temperature or the low-temperature part of the system, is a complicated and quite annoying problem. The increase of polyvinyl chloride in MSW is the main reason for high-temperature corrosion in the boiler, but there are also other factors which can directly or indirectly (catalytic process) cause damage on tubes in the high-temperature zones of the boiler. The corrosion can be partly explained by the following process.

On the superheater pipes, especially on tubes directly exposed to the hot gases, both sulphates and alkali chlorides are condensed. These react with sulphur oxides releasing chlorine or chlorides, which further react with SiO_2, Al_2O_3 and Fe_2O_3 at the temperature of incrustment on the tubewall. This reaction can be in balance as long as there is sufficient matter in the incrustment on the tube wall to react with the chlorine and chlorides being constantly released. If the amount of free chlorine exceeds a certain value, which depends on the composition of the incrustments, it reacts with the oxide layer of the steel or directly with the steel itself and corrosion is initiated.

Low-temperature corrosion occurs when the temperature falls below the dew point of the flue gases. A high amount of sulphur (not in MSW) and/or chlorine lowers the dew point and therefore it is advisable not to let the temperature fall below 250°C in the main flue gas stream to avoid local reduction of the temperature below the dew point. By water washing the gases to bring down the amount of HCl, HF and eventually the heavy metals, the dew point is reached. Accordingly, one has to use quite sophisticated materials or rubber-coated common steel, wherever the temperature allows.

The three most important factors (except investment costs) affecting the result of the incineration with heat reclamation of MSW are

availability,
steam production rate, and
amount of ashes, clinker and untreated refuse for disposal.

The availability is defined as the true running hours compared with the planned running hours. It has proved difficult to bring availability to a figure over 80%, though there are a few plants with an availability of over 90%.

The steam production rate (tonnes steam per tonne MSW) is of course influenced by the availability by making a statistical calculation for a whole operating cycle (1 year), and the same goes for the amount of waste which has to be dumped.

Fig. 11.3 shows the result from eight years of MSW incineration in Nürnberg, West Germany. Comparison of the lower curve (ashes and clinker) with the upper one shows a much higher proportion (over 40%) of the amount to be dumped to the amount of MSW hauled to the incinerating plants than is commonly thought to be the case (about 30%).

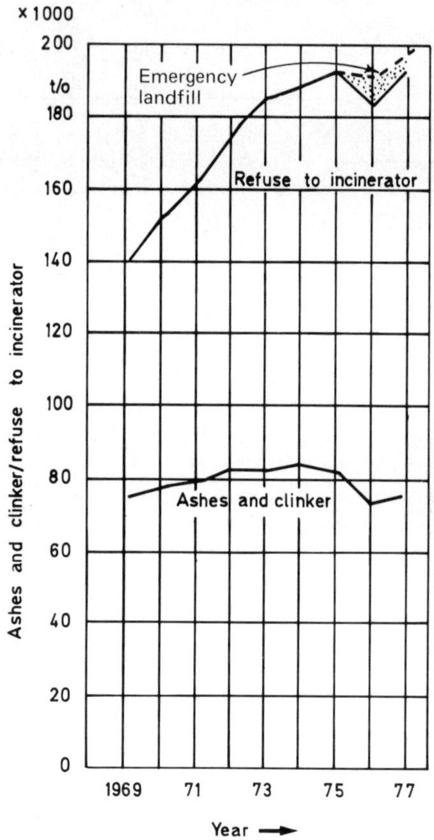

Fig. 11.3 Incinerated MSW. Ashes and clinker to deposit. (Reproduced, with permission, from R. Pohle, *VGB Kraftwerkstechnik* **59**, 2 (1979).)

Fig. 11.4 shows the steam production rate at the Nürnberg plant. Bearing in mind that the common figure for oil-produced steam is 14:1, the result indicates that the 'rule of the thumb' that 5 tonne of MSW equals 1 tonne of fuel oil is in this case quite optimistic. Occasionally, the figure can of course be even better than 5:1, but allowing for availability one gets a substantially poorer result overall.

The beneficial effect of selling steam (and electricity) at a fair market price is badly shaken by this double deficit; less steam to sell and more untreated MSW to dump. In this connection it is fully justifiable to point out the convenience of

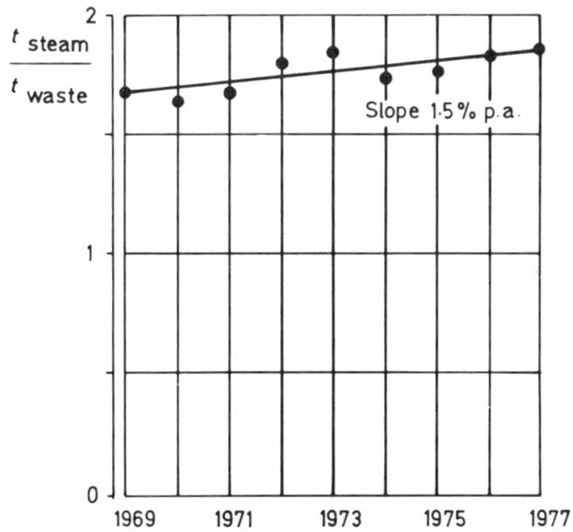

Fig. 11.4 Steam production rate. MVA Nürnberg. (Reproduced, with permission, from H. Knoll, *VGB Kraftwerkstechnik* **59**, 2 (1979).)

having more than one unit installed or to apply an availability factor not more than 60% to a serious calculation.

PRODUCTION AND UTILIZATION OF REFUSE-DERIVED FUEL (RDF)

Parallel to the development of different methods of controlled gasification of wastes (including pyrolysis) a lot of work has been put into research and tests in the field of RDF-burning, i.e. more or less complete transformation of MSW into combustible pellets. Though the outcome of the tests has not been totally successful, there are reasons to believe that, at least in heavily industrialized countries, the burning of RDF will take the place of incineration within the next 10 years. The main reason for this is the increased demand for recycling materials in an unoxidized form together with the creation of some sort of compost. There is a lot of sewage sludge, digested and dried, waiting to be utilized and the organic matter in the MSW can act as a carbon-rich partner to the nitrogen in the sludge. Papers and journals in MSW can be sorted out through household collection (source separation) and thus will not affect the result of a chosen method in comparison to that of another; but if the feasibility and economy of source separation are not accepted, and as a consequence mechanical separation is accepted, the part that will be left for burning automatically achieves the shape of an RDF of some degree. Together with pulverized or shredded coarse refuse from households, shops, offices, gardens and building activities, these can be transformed into quite valuable RDF. Apart from such RDF produced as a by-product, there are advantages giving priority to the fabrication of RDF:

(i) a fuel with high heating value (15 – 20 MJ/kg),
(ii) a fuel with lower content of heavy metals and, depending on method, lower content of PVC than untreated MSW,
(iii) a product relatively easy to transport and store.

These are the main inducements to activate innovation in this field. Where coal-fired boilers are prevalent, conversion to firing with a combination of coal and RDF can be tested and established without too much resistance from the market. In countries like Sweden with practically no coalfiring, but with bark-fired boilers without moving grates, the introduction of RDF is subject to resistance from the overall economy and a demand to prove its usefulness in common bark-fired boilers. What is lost in RDF processing must be regained in simpler boiler installation.

The various grades and qualities of RDF are illustrated by reference to six methods of production of increasing complexity and hence cost.

RDF 1 = untreated MSW (i.e. a 'false' RDF)
RDF 2 = pulverized MSW
RDF 3 = pulverized and magnetically separated MSW
RDF 4 = pulverized, magnetically separated, screened MSW
RDF 5 = pulverized, magnetically separated, screened dried, pelletized MSW
RDF 6 = pulverized, magnetically separated, screened, dried, pulverized (to coarse dust) MSW

Refuse Sorting Processes

There is in Sweden a system (the BRINI-system) which has been described in the literature. Fig. 11.5 gives a good view of the material disposition of the system, and as a comparison to BRINI-pellets a Sankey-diagram for the American ECO-FUEL II product is presented in Fig. 11.6. The heating value of the product is about 16 MJ/kg for BRINI-pellets and about 18 MJ/kg for ECO-FUEL II. Seen as a total, a greater amount of heat/tonne fed waste is achieved in the former case, but ECO-FUEL II is delivered as a powder, which is supposed to be burnt in coal-fired boilers. On the other hand, BRINI-pellets require boilers with moving bottom grates or similar.

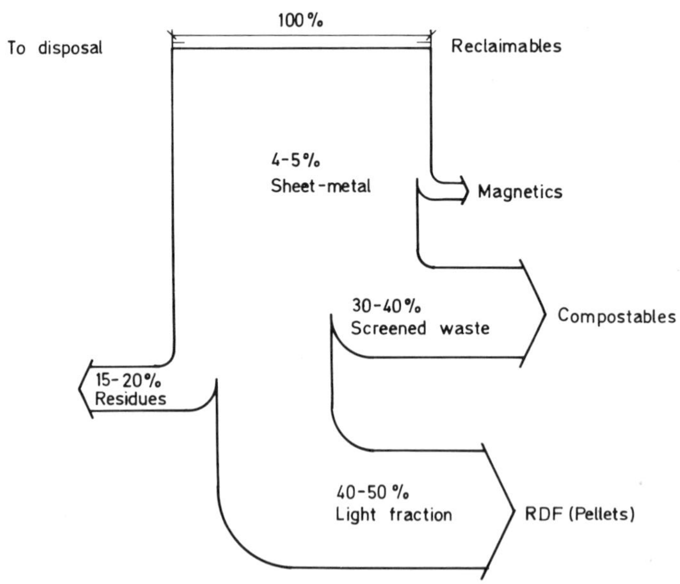

Fig. 11.5 Sankey-diagram for Brini-pellets. (PLM Återvinning AB, Malmö).

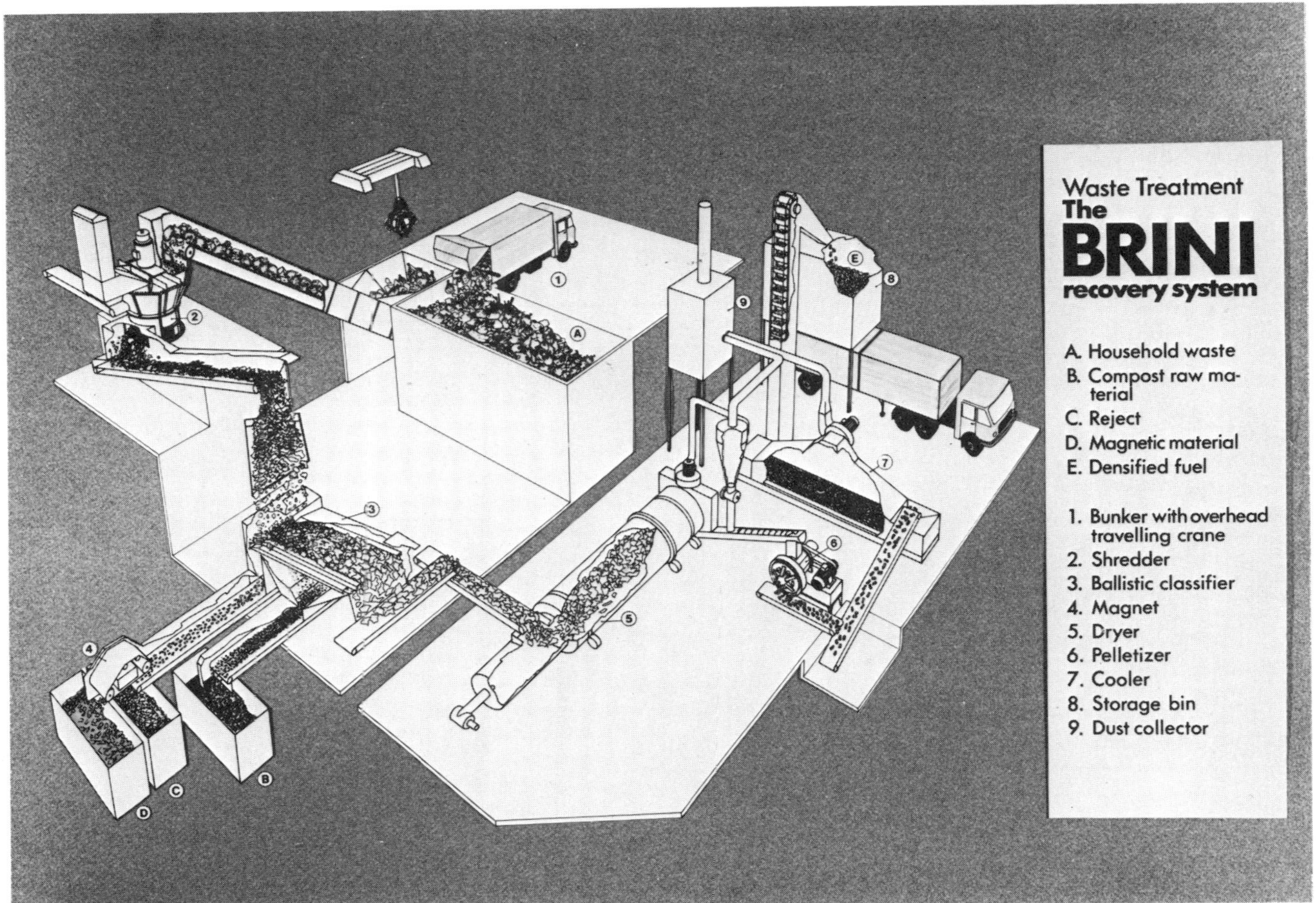
Waste Treatment
The
BRINI
recovery system
A. Household waste
B. Compost raw material
C. Reject
D. Magnetic material
E. Densified fuel
1. Bunker with overhead travelling crane
2. Shredder
3. Ballistic classifier
4. Magnet
5. Dryer
6. Pelletizer
7. Cooler
8. Storage bin
9. Dust collector

Pellets from the Brini recovery system.

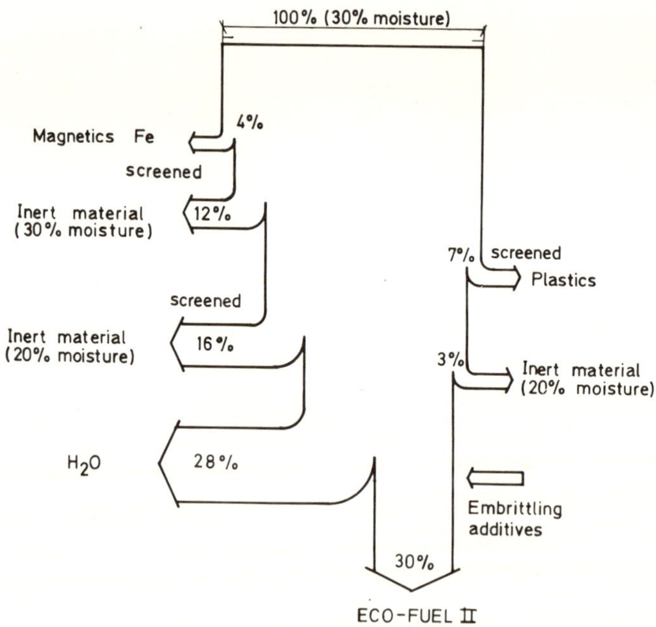

Fig. 11.6. Sankey-diagram for ECO-FUEL II. (Combustion Equipment Associates, Inc., New York).

Table 11.2 shows the product specification of BRINI fuel.

TABLE 11.2 Product Specification: Brini Pellets

Composition (dry product) (Separated from normal solid household waste)	
Paper	approx. 70% by weight
Plastic foil	approx. 10% by weight
Other combustible materials	approx. 16% by weight
Other non-combustible products	approx. 4% by weight
Sulphur content	approx. 0.2% by weight
Bulk density	approx. 450 kg/m^3
Size	32 times 32 mm square (length: about 30 mm)

Fig. 11.7 is a flowchart of the process. As can be seen, the process can easily be divided into two separate modules, the first one producing RDF 4 and the second one RDF 5.

The appendix to this chapter contains data from a full-scale test with up to 90% (in heating energy) of BRINI and the rest bark and saw dust in an ordinary bark boiler in Landsbro in the south of Sweden (April 1979). Certain conclusions can be drawn from these tables. There is an obvious reduction of chimney emissions of particle-bonded mercury, total acid and HCl. On the other

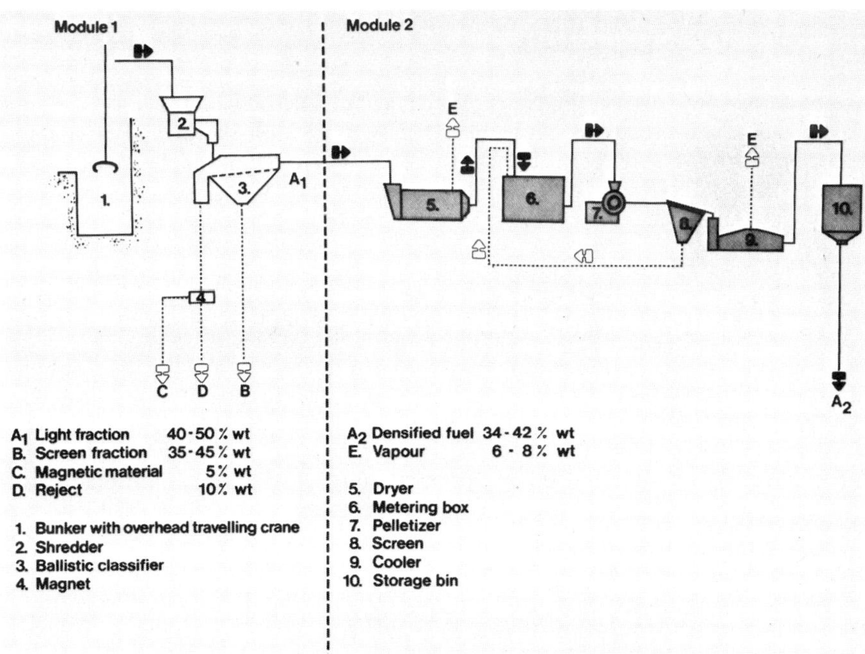

Fig. 11.7 Refuse-derived fuel (RDF) — The PLM Brini system.

hand, emissions of gaseous mercury and of mercury in slag and separated substances are equivalent to those of three large-size waste incineration plants, whereas emissions of other heavy metals are 10 – 50 times lower. It is difficult to explain why the mercy contents in Brini pellets is not lower, since the removal of batteries, electrical devices, instrumentation products, etc., should entail a more substantial reduction. The tables show a difference in the reported quantity of mercury for Brini v.s waste (Laxå, Sweden and Sintef, Norway, respectively) which is as little as 50%. American investigations give a range of 1 – 3 g/ton, and the values used are likely to be detrimental to the Brini method as regards mercury contamination.

The high fire box temperature during the Landsbro test was probably favourable for POM emissions but unfavourable for the evaporation of mercury, i.e. the larger part of the metals are discharged in the gaseous emission and a lesser part is bonded in ashes and slag. Also the NO_x share increases as the temperature rises. The fact that the figure representing the contents of mercury in ashes, slag and dust are not correlated with this statement can be attributed to an accidental occurrence during the sampling.

The combustion process was strikingly smooth in the boiler fireplace, a fact which is reflected by, for example, the low dust contents in the raw gas. No increase in dust pollution, compared with that occurring with wood fuel exclusively, was registered.

The quantity of combustible components in ashes and slag when heating with 90% of Brini-pellets was approximately 1%, i.e. considerably lower than what is usual in conventional waste incineration in minor plants.

Summing up, it can be concluded that from the emission aspect the Brini pellets occupy an intermediate position between municipal waste and wood fuel, with a distinct shift towards those values that are assumed to be valid for wood fuels.

The Costs of Producing RDF

As has been stated above, the costs of producing RDF rise with the improvement of RDF as a substitute fuel. Fig. 11.8 shows the result of calculations of the 'fabrication' costs for different qualities of RDF (identified as RDF 1, 2, 3, etc., earlier). It assumes that all costs are allocated to the RDF product. This accounts for the steep slope of the curve between RDF 3 and 5 as the costs rises simultaneously to a decrease in the amount of usable product. This might be rather unfair to the more sophisticated methods, but in order to eliminate the negative effect, a curve showing the alternative costs for the municipality must also be drawn, also referring to the usable amount of RDF at each stage.

Including costs for disposal of clinker and ashes, the extra value of RDF 5 (in this case Brini-pellets) is about 120 SKr ton as delivered at the MSW-handling plant, i.e. about 50 SKr/MW_{th}h.

The ultimate question is, of course, whether it is realistic for an industry or a central heating plan to pay such a price for this alternative source of energy. Figs. 11.9 and 11.10 illustrate the results from a rather laborious calculation of the costs involved.

Returning to the example above and assuming a quantity of saleable RDF corresponding to an output of about 20 MW_{th} at the consumer, including all losses, the costs will not exceed 20 SKr/MW_{th}h excluding costs for RDF delivered at the plant. Assuming a transportation distance of about 100 km, the RDF can be delivered at a price of about 52 SKr/MW_{th}h, which gives an

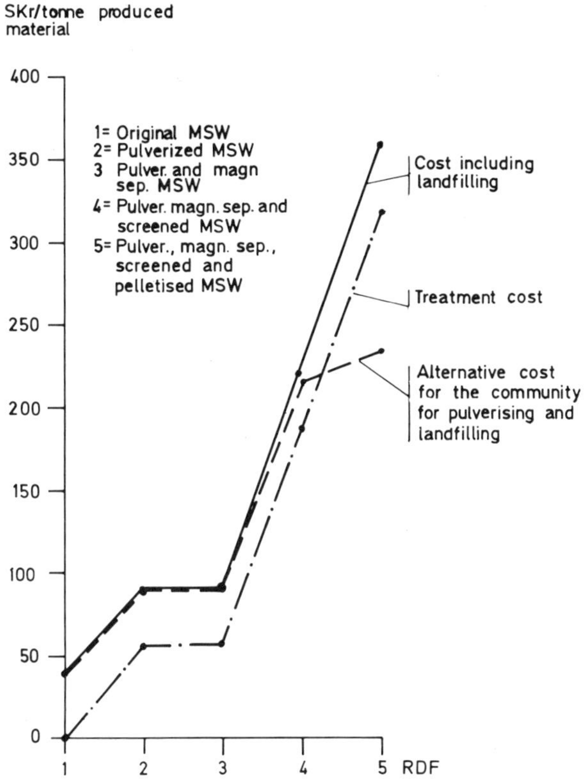

Fig. 11.8. Costs of producing different kinds of RDF.

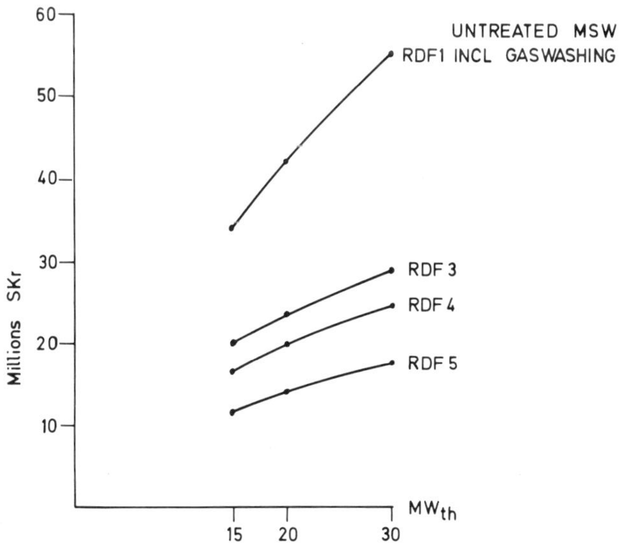

Fig. 11.9. Capital costs of plant erection as a function of yearly throughput of RDF expressed as thermal effect.

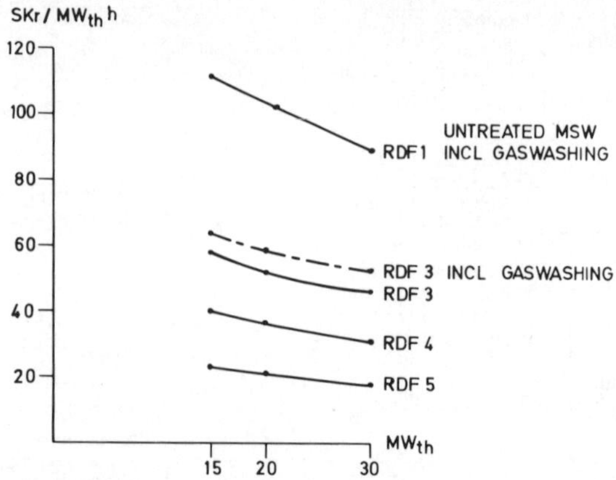

Fig. 11.10. Running costs per produced $MW_{th}h$ as a function of yearly throughput expressed as thermal effect.

overall cost of about 72 SKr/$MW_{th}h$. If the consumer is forced to install some sort of flue gas washer, the costs will rise to about 85 SKr/$MW_{th}h$, which, compared to that for oil, is a fairly acceptable price.

Refuse Burning

Tests have been carried out in Hamburg by burning RDF 2 and 3 in the ordinary mass-incineration plant. It was found that the resulting steamflow was much more steady than the usual one, but the dust load became higher and the quality of ashes and clinker was not as good as was expected. It was also found that the old model of grate, i.e. three separate moving grates with steps in between, was *not* suitable for burning RDF 2 or 3.

In Eksjö in Sweden a full-scale test was run with a 5 MW_{th} fluidized bedreactor (see Fig. 11.11). The results from the start-up tests are not yet officially released, but there seems to be an opinion that the feeding of pulverized MSW over a fluidized bed gives certain problems from dust carrying over. A suggestion to feed the RDF 3 under the surface of the bed has been rejected because of the risk of backfire in the feeding pipe or screw. The requirement is for MSW in the form of small blocks which will remain in the bed like coal or wood chips. Tests have been performed successfully with bark and chips, but to fire coal the bed has to be filled with cooling surfaces. The difficulties with start-ups and regulation has in this case been solved by dividing the bed into three separate chambers, the centre one formed (above the grid) by unburned fuel which piles up around the walls.

This spring, a new RDF 3-fired boiler (see Fig. 11.12) is due for commissioning. This boiler is a modified bark-boiler with a rating of 18 MW_{th} when fueled with MSW. The flames are sucked downwards through the upper stationary grid and then sideways to a pre-combustion chamber where the oil-burners (50 MW_{th}) are situated. The moving grate is of the Kablitz type and its purpose is to discharge the ashes and clinker. This unit is not especially cheap, as it is a prototype which has been promoted with help of government subsidiaries, but when in production the costs can be brought down at least to

The first shredded-waste boiler of fluidized bed type in the world, the Eksjö 5 MW_{th}-fired hot water boiler.

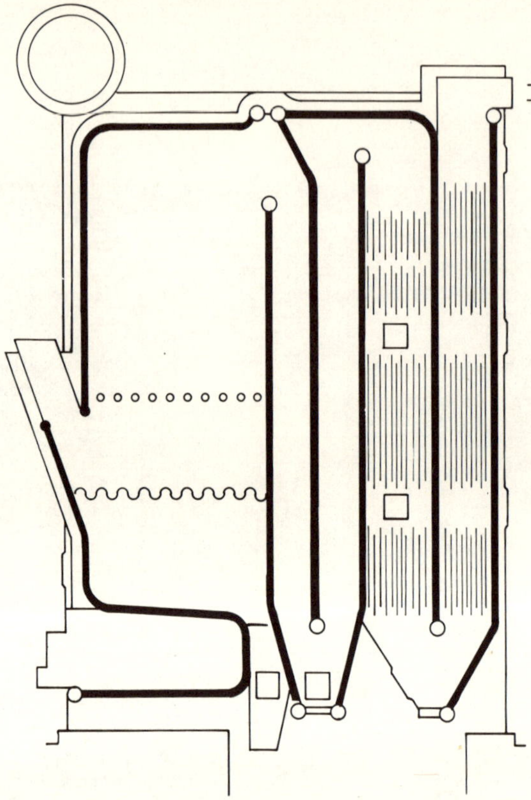

Fig. 11.11 5-MW_{th} MSW-fired heating water boiler with fluidized bed. (Generator Industrial AB, Partille, Sweden).

the level of fluidized bed units. In the comparative costs diagram, this has been taken into account.

Data on impurity levels in a wide range of solid fuels and pollutant emission on combustion are given in the appendix.

PYROLYSIS

There are a few drawbacks connected to the methods of incineration: high costs, noxious gases and the need to use the main product, i.e. thermal energy, on the spot. Many firms and institutes have investigated the possibility of pyrolyzing (gasifying the organic matter in the absence of oxygen) the waste. Investigations can be divided into two categories: high-temperature pyrolysis, and medium- and low-temperature pyrolysis. The principles of the thermal processes are explained in Tables 11.3 and 11.4.

The main difference is that in the lower temperature processes more tar, char and oil products are formed, which can be recovered, whereas at higher temperatures molten slag is formed. High-temperature pyrolysis (better known as gasification) is quasipyrolysis process and differs from the conventional mass burning process in that the burning of volatiles is carried out in a separate section

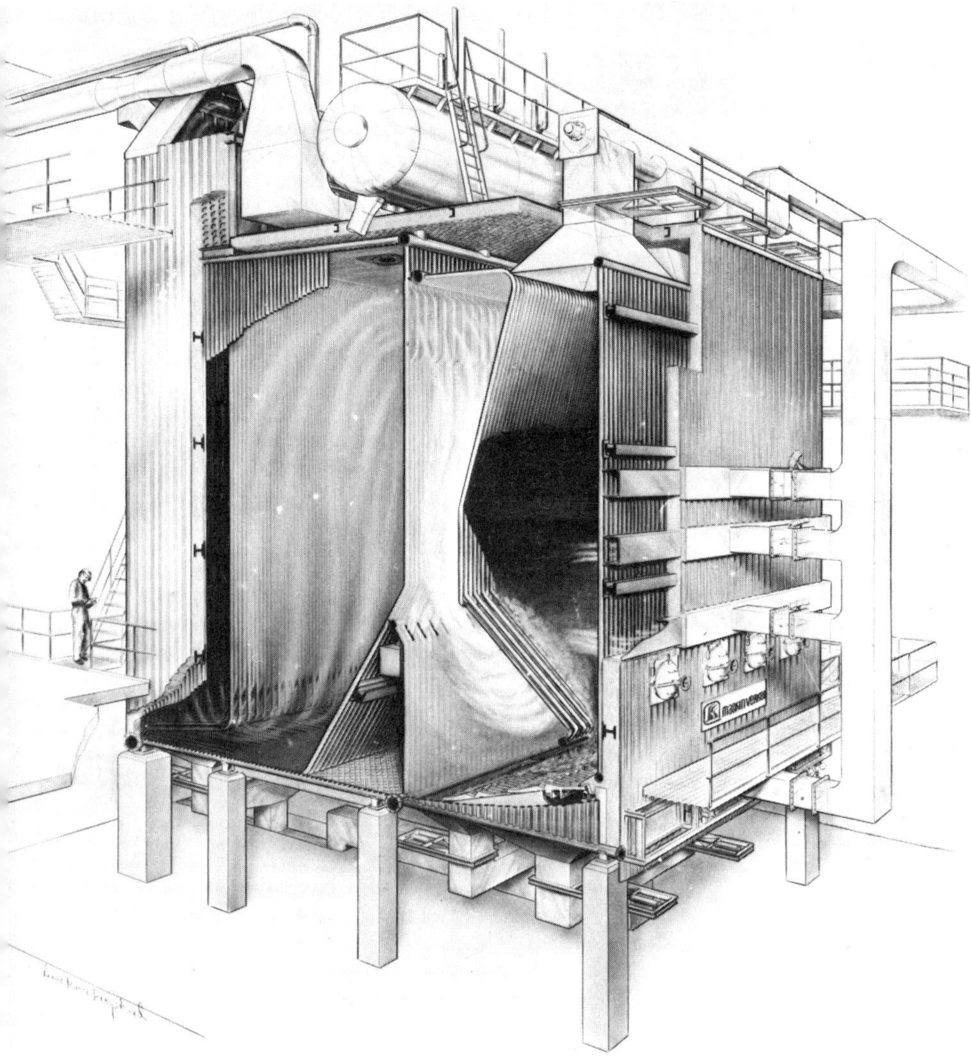

Fig. 11.12 Modified bark-boiler in Borlänge, Sweden (Götaverken Sv. Maskinverken, Kallhäll, Sweden.)

at such a high temperature that the solid products (ashes) can be drawn out as molten slag and brought together with that of the main reactor.

The theory for the process of low-temperature pyrolysis in that of the real pyrolysis process is that no oxygen is supplied to the process. The amount of flue gases generated is much less than that generated by incineration and the flue gas pattern is also different. Table 11.5 gives a comparison between the principal outlets from the two different plants. It can be seen that the chlorides are to be found in the ashes and clinker in the ideal pyrolysis process in the form of NH_4Cl, which will cause no damage to the ground water, and no NO_x is generated (with the exception of oxidation of nitrogen by oxygen present in the

TABLE 11.3 Simplified Picture of the High-Temperature Pyrolysis Process

Progressive stages in decomposition	Temperature (°C)	Products
Drying	20 – 100	Steam with odorants
Heating	100 – 200	
Pyrolysis	200 – 500	The formation of long-chain hydrocarbons such as tar, oil, organic acids, phenol; fixed carbon; small amounts of permanent gases
Formation of gases	500 – 1200	Decomposition of the long-chained hydrocarbons to H_2 CO, CH_4 and carbon. The transformation of carbon and steam into H_2 and CO; steam reforming and shift reactions
Melting of inorganic residues	1200 – 1700	Melting of the inorganic residues to slag; some reduction of metal oxides

TABLE 11.4 Thermal Degradation Processes during Pyrolysis of Organic Matter in the Low and Medium Temperature Range

Temperature range (°C)	Main process
100 – 120	Drying — removal of free water; no visible degradation
120 – 250	Deoxidation and desulphurization; visible degradation, decomposition of cell-bound water and cellulose
above 250	Depolymerization; growing decomposition of H_2S
around 340	Breakdown of alimatic bonds; methane and other hydrocarbons start to be released
around 380	Carbonizing (carbon enrichment of pyrolysis products)
around 400	Breakdown of bonds of the carbon-oxygen and carbon-nitrogen chains
400 – 420	Transformation of bitumen into pyrolysis-oil and pyrolysis-tar
420 – 600	Cracking of bitumen into more stable compounds (including short-chain hydrocarbon and graphite)
above 600	Formation of aromatics

charged waste depending on the highest temperature in the reactor). In spite of many trials and a profoundly sound theoretical basis, pyrolysis has not so far proved its capability of full-scale production. The only full-scale system on the waste handling market is that of the Andco Torrax high-temperature gasifying system, which has been applied to four different plants on the European continent. After many adjustments to the first plant, the system appears to stand a good chance to be fully commercial. Unfortunately, the promising results in early reports turned out to be too optimistic and either the method (and this goes for pyrolysis as a whole) will disappear from the scene of waste handling rivalry or its exponents will have to come up with quite a new approach to the problems.

TABLE 11.5

Chemicals	Incineration		Pyrolysis	
	In the clinker	In the gas	In the clinker	In the gas
System-bounded toxic chemicals				
F	Fluoride	HF	Fluoride	HF
Cl	Chloride	HCl	NH_4Cl	HCl
S	Sulphate	SO_2*, SO_3*	Sulphide	H_2S*
N	—	NO_x, N_2	NH_3, NH_4Cl	NH_3*
Metals	Metal oxide		Metals (reduced)	
Process-generated toxic chemicals				
C	Incompletely burned organic material		Tar, Phenols, Cyanide	Tar, Phenols, Cyanide

*Can be washed out by water scrubbing.

APPENDIX: IMPURITIES IN VARIOUS SOLID FUELS AND FROM THEIR COMBUSTION

TABLE 11.6 Metal Contents of Different Fuels and Wastes, in Terms of ppm or g/ton

Values for bark not reported.

	Brini	Waste		Coal	Heating oil	Peat	Wood
		Laxa, Sweden	Sintef Norway				
Cd	2	3.5	4.4	0.01 – 2	0.01	0.17	0.06 – 1.70
Co	2	4.0	6.0	1 – 40	0.2 – 0.5	0.25	0.05 – 0.30
Cr	28	28	130	2 – 50	0.02 – 0.05	1.1 – 5.0	0.19 – 2.30
Cu	60	400	110	0.04 – 40	0.1 – 0.3	2.7 – 11.6	1.9 – 23.0
Hg	1.1	1.4	1.5	0.07 – 1.5	0.002 –	0.05	0.01 – 0.07
Mn	239	880	170	25 – 100	0.03 – 0.1	10 – 156	5 – 194
Ni	4	15	19	4 – 60	6 – 23	1.0 – 2.4	0.3 – 1.4
Pb	78	210	160	2 – 30	0.07 – 2	1.0 – 19	0.52 – 4.6
Zn	510	1200	840	5 – 100	0.3 – 1	6 – 53	7.0 – 92
Sn	20						
As	1.5						
F^- (fluoride)	6						
F (fluorine)	91						
S (wt%)	0.2						
Cl^- (chloride) (wt%)	0.2						
Cl (chlorine) (wt%)	0.4						
Fe (wt%)	0.1						
Al (wt%)	1.1						

TABLE 11.7 Metal Contents per Energy Unit in Different Raw Materials (tonne/$MW_{th}h$)

Shale 1, from Ranstad, Sweden, is used as uranium raw material for light water reaction. Shale 2 refers to fossil fuels in Ranstad shale. Wood and 'Energy forest' are Biomass.

Metal		Coal	Heating oil	Biomass	Peat
Arsenic	As	0.7	0.007	—	
Cadmium	Cd	0.08	0.0004	0.01–0.3	0.03
Cobalt	Co	0.8		0.01–0.1	0.04
Chromium	Cr	1.6	0.002	0.04–0.4	0.2–0.9
Copper	Cu	2		0.4–4	0.5–2.1
Mercury	Hg	0.013	0.0003	0.002–0.01	0.009
Manganese	Mn	5		1–30	2–30
Molybdenum	Mo	4	0.02		0.8–
Nickel	Ni	2	1	0.06–0.3	0.2–0.6
Lead	Pb	1.3	0.01	0.1–0.8	0.2–3.4
Vanadium	Vd	3	2		—
Zinc	Zn	6		1–17	1–10
Energy content $TW_{th}h$/million tonnes		7.5	11.5	2.7	2.4

	Shale 1	Shale 2	Brini	Wood	'Energy forest'
	4.6	18	0.4	—	—
	0.27	1.0	0.5	0.011–0.06	0.08–0.3
			0.4	0.000–0.06	0.009–0.03
	39	150	6.7	0.035–0.4	0.04–0.3
	13	52	14.4	0.4–1.0	0.9–4.3
	0.04	0.15	0.3	0.002–0.004	0.005–0.01
	30	120	62	2.1–33.5	0.9–8.1
	41	160	—	—	—
	24	95	1.0	0.056–0.21	0.09–0.3
	1.7	6.7	18.7	0.09–0.85	0.01–0.3
				—	—
	16	62	122	0.9–10.4	8.3–17
Mean value	8.2	2.1	4.2	5.4–5.8	5.4–

TABLE 11.8 Sulphur Contents in Certain Solid Fuels

	Tonnes of sulphur per $TW_{th}h$
Coal	890–4400
Heating oil	430–3500
Peat	0–4200
Biomass	70–370
Shale	28 000*
Wood	110
Bark	Not known
Brini	480

*fossil energy in Ranstad shale.

TABLE 11.9 Chimney Emissions from Refuse Incineration Plants and from Combustion of Brini Pellets per $MW_{th}h$

		Sävenäs[1] (Gothenburg, Sweden)	Sjölunda[1] (Malmö, Sweden)	Högdalen[1] (Stockholm, Sweden)	Brini[5] raw gases (Landsbro, Sweden)	Brini[2] gases from chimney (Landsbro, Sweden)
Hg	g	0.060	0.180	0.032	0.114	0.006
Cd	g	0.522	0.180	0.450	0.761	0.72
Pb	g	4.96	3.3	7.7	15.9	19.1
Zn	g	11.76	8.3	?	64.54	76.7
Mn	g	(0.869)	?	0.354	4.16	0.582
Cu	g	(0.608)	?	1.359	1.74	0.970
Cr	g	(0.444)	0.207	0.189	35.7	1.300
Ni	g	(0.222)	?	0.414	24.6	1.040
F	g	(1.737)	?	?	?	?
$SO_2 + SO_3$	kg	0.359	0.443	3.40[3]		0.55
HCl	kg	1.955	1.330	2.273		0.127
HF	kg	0.009	0.009	0.0005		0.006
Total acid	$m/kW_{th}h$	25 – 175	?	?		32.4[4]
No_x	kg	0.638	1.182	0.044		0.78
Hg, gaseous	g	0.2	0.1	0.5		0.15

[1]Electric filter. Calorific value 2.6 $MW_{th}h$/ton.
[2]Multi-cyclone. Calorific value 3.6 $MW_{th}h$/ton.
[3]Probably 2 – 3 times too high.
[4]Estimated value.
[5]10% Wood, 90% Brine.

TABLE 11.10 Chimney Emissions from Heating with Solid Fuels Calculated per $MW_{th}h$

Bark values not reported.

		Coal	Wooden fuel*	Wood (nominal values)	Peat	Brini**
Hg	g	0.014	0.03	0.012	0.002	0.006
Cd	g	0.008	0.19	0.20	0.001	0.72
Pb	g	0.053	6.72	0.30	0.006 – 0.088	19.1
Zn	g	0.12	28.2	6.4	0.03 – 0.32	76.7
Mn	g	0.026	1.48	5.6	0.04 – 0.70	0.582
Cu	g	0.01	0.54	0.6	0.02 – 0.05	0.970
Cr	g	0.019	1.97	0.1	0.006 – 0.025	1.300
Ni	g	0.009	1.08	0.1	0.006 – 0.020	1.040
$SO_2 + SO_3$	kg	1.78 at 0.3%S		0.1	0.22 – 2.07	0.55 as SO_4
HCl	kg	—		—		0.127
HF	kg	—		—		0.006
Total acid	kg	—		—	0.42 (estimated)	(g 32.4 (nominal)
NO_x	kg	0.89 as NO_2		2.2	2.22	0.78 as NO_2
Hg. gaseous	g	12				0.00015

*Bark + chips (Landsbro, Sweden). Calorific value 2.23 MWh/tonne.
**10% Wood + 90% Brini.

TABLE 11.11 Metal Content in Ash and Slag Including Separated Dust when Heating with Waste and Wood Fuel per $MW_{th}h$

Bark values not reported. Values in parenthesis are estimated. All values are in grams.

	Sävenäs (Gothenburg, Sweden)	Högdalen (Stockholm, Sweden)	Brini* (Landsbro, Sweden)	Wood fuel—chips + bark (Landsbro, Sweden)
Hg	0.077	0.116	0.104	0.002
Cd	(4.44)	3.72	0.114	0.053
Pb	795	326	8.69	3.48
Zn	1 295	609	23.1	7.07
Mn	(14.2)	259	47.8	26.2
Cu	363	177	35.4	15.2
Cr	(6.66)	8.08	1.57	0.92
Ni	(6.66)	17.82	1.87	0.42

**10% Wood, 90% Brini.

Bengt Enhörning has worked as a consultant since 1951, with the exception of four years (1957-1962) spent at the government company Atom Energy in Studsvik. He was formerly engaged in the planning of steam-raising plants and heavier piping installations. He is now an authorized inspector of pressure vessels in Swedish nuclear plants and a specialist in the re-use of energy and materials from waste.

12 The Technology and Economics of Thermal Processing of Refuse

A. V. Bridgwater

INTRODUCTION

Refuse consists of a heterogeneous and widely variable mixture of waste materials which include paper, vegetable and animal wastes (putrescibles), rag and textiles, plastics, metal, glass, dust and cinder, and water. Although compositions vary widely according to time of year and location, Table 12.1 gives typical European figures for reference.

TABLE 12.1 Typical European Refuse Composition

	Dry basis (%)	Wet basis (%)
Paper	39	27
Vegetable and animal wastes	21	15
Rag and textiles	3	2
Plastics	4	3
Metal	8	6
Glass	9	6
Dust and cinder	16	11
Water	—	30
	100	100
Heat energy content	10 000 kJ/kg	7000 kJ/kg

Typically, therefore, over half the contents of refuse on a dry basis have a high energy content. Unless controlled, this energy is utilized biologically to create an objectionable environment. There have been many attempts over a very long period of time to control utilization of this energy either to avoid nuisance or for exploitation in a more positive way. Table 12.2 summarizes the alternatives and offers brief comments on their relative technical and economic merits. In most cases the objective is profitable utilization of the energy in some form rather than a satisfactory method of disposal. This point of view has been enhanced by the energy crises of the past decade, and for refuse it is put into perspective by an estimate that the energy content of the 18 million tonnes per year of refuse in the UK represents nearly 5% of the annual energy demand; or it could supply about 2½% of this demand.

TABLE 12.2 Energy Recovery from Refuse

Note: Front or, less likely, back end resource recovery options may be included for any option. All thermal processes give high-volume reduction and a sterile product.

Method	Advantages	Disadvantages
Incineration		
(a) with heat recovery	Good method for district heating Higher burn-out efficiencies can be expected with prepared fuel (RDF) than with unprepared refuse Commercial available plant Can be developed to air conditioning system	Corrosion of boiler tubes at high steam temperatures Steam flow not sufficiently dependable to run power plant auxiliary systems High initial costs Slagging of heat exchange surface can give high cleaning costs and downtime Pollution problems Reliability of supply doubtful Market demand variable
(b) with electricity generation	Total electric power production package available Good overall system efficiency	Serious technical problems with gas clean-up before turbine New electrical generation equipment required Very high initial and running costs Other problems as (a)
Gasification		
To give gas and ash	Produces low to medium heating value gas Feed preparation not essential, although preferred Existing power plant can be used High overall system efficiency Higher-value products than with incineration Fuel gas usable in most boiler types Technology more advanced than pyrolysis Gas may be employed as chemical feedstock	Potential plugging of slag Fuel gas not compatible with natural gas without additional processing/expenditure Storage of fuel not viable High initial and operating cost Unproven viability Waste-water disposal problem Low heating value of gas necessitates local use Tars need recycling
Pyrolysis		
To give oil, gas and char	Oil can be used in conventional boiler with minor modifications Existing power plant can be used Higher-value products than incineration Feed preparation not essential, although preferred	Technology unproven Problems with corrosiveness and storability of pyrolytic oil High initial and operating costs Costly feed preparation Waste-water disposal problem Utilization of all products necessary
Liquefaction	Produces high calorific value oil Product composition can be controlled Potentially economically attractive	Still at early development stage High pressure operation Special charging and discharging problems

Other Methods		
(a) Solid fuel preparation as RDF/dRDF*	Existing facilities can be used with minor modification to generate steam or electricity High overall system efficiency Relatively low costs dRDF improves storage and handling Largely proven technology Plant available commercially Compatible with solid fuels	Low bulk density of unprepared refuse makes storage difficult Potential increase in particulate loading and pollution Densifying/pelletizing equipment still presents problems High costs and unproven viability
(b) Anaerobic digestion to give methane	Existing steam or electricty generation plant can be used Product compatible with SNG after carbon dioxide removal	Sensitive to moisture and oxygen environment Very low overall system efficiency Product contaminated with carbon dioxide which requires separation Reaction rates very low, requiring large reactors and long residence times Residue disposal problem Viability doubtful on a large scale
(c) Fermentation to chemicals	Technology well developed High-value products recovered	Sensitive to contamination High energy costs in purification from an aqueous base High costs Residue disposal problem Viability doubtful

*RDF = refuse derived fuel; dRDF = densified RDF.

THERMAL PROCESSING

Waste materials may be thermally processed in a variety of methods which are here characterized by the level of oxygen present:

1. Incineration, where complete combustion or oxidation occurs, often with excess air, resulting in fully oxidized products.
2. Gasification, where incomplete combustion or oxidation occurs, using either air, oxygen-enriched air, or oxygen. Only a fraction of the material burns, thereby generating sufficient heat to thermally decompose the rest of the material. The products are not fully oxidized and have a heating value. Other reagents such as steam (water) or carbon dioxide may be added.
3. Pyrolysis, where no combustion or oxidation takes place, but thermal degradation of the material is effected by adding heat directly or indirectly. Again, reagents or catalysts may be added to increase yields of preferred products. Included here are liquefaction processes which are high-pressure liquid-phase processes.

This is only one way of classifying thermal processes and is a considerable simplification, as pyrolysis occurs in gasification, and both pyrolysis and gasification occur as intermediate steps in incineration. In addition, there are many other reactions or processes occurring, such as hydrogasification, methanation, steam reforming, partial oxidation and shift, the extent of which vary according to the above classification and a wide range of reaction variables.

In all cases the energy content of the waste is utilized to give heat, or a product with a heating value. This product can be gas, liquid and/or solid, which is referred to as a primary product, and these can be subsequently converted and upgraded to secondary fuels or chemical raw materials. The range and interaction of alternatives is outlined in Fig. 12.1.

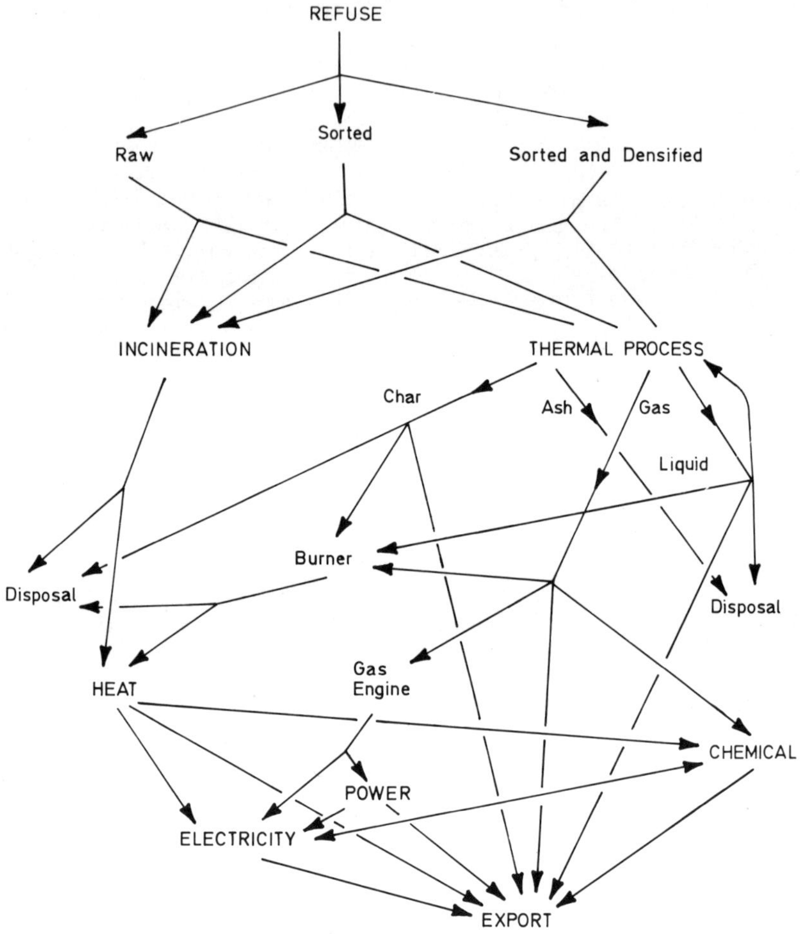

Fig. 12.1 Products of incineration and pyrolysis.

UTILIZATION OF PRODUCTS

The range of products and how they can be further converted to a more useful form is shown in Fig. 12.1. The energy may appear in thermal, mechanical or electrical form, which has to be used as it is produced — i.e. it cannot be stored. Alternatively, it may be produced in chemical form, as a fuel or raw materials, which may be stored and used as required.

In chemical form, the products always include gas, containing carbon monoxide, hydrogen and some methane; some liquid, containing a wide range of oxygenated organic compounds up to pitch-like materials, as well as water which will dissolve some of the lower boiling point chemicals thereby creating a disposal problem; and a solid residue which will contain carbon or char if relatively low reaction temperatures (below around 900 °C) are used, as well as ash and any other non-reactive materials such as glass and metal.

The gas also contains carbon dioxide, and nitrogen if air has been used for gasification. It may be used as a fuel gas directly when the process becomes a two stage incinerator (e.g. Andco Torrax Process) or piped to a local/distant facility. Low heating value gas (3—5 MJ/N m^3) by air gasification has to be burned immediately to conserve the sensible heat. Oxygen-enriched air gasification products, oxygen gasification products and pyrolysis gas products of 12 – 20 MJ/N m^3 may be cleaned and piped to a local user. The maximum economic distance increases as the heating value increases, but is unlikely to exceed about 10 miles (15 km).

After cleaning and carbon dioxide removal, the gas becomes synthesis gas or 'syngas' which, in principle, is suitable for methanol synthesis, methane synthesis, synthetic crude oil production or hydrogen production (see Fig. 12.2). Contamination with nitrogen reduces efficiency and increases costs to an unacceptable level and thus air gasification products are not satisfactory for this purpose.

The liquid product consists of four categories: immiscible organic compounds of lower density than water; water; organic compounds miscible with water; and immiscible organic compounds of higher density than water. The extent of organic compound formation is dependent on a wide range of factors whose effect is not well understood. Some processes have sought to maximize the 'oil' yield (e.g. Garrett/Occidental process) while others have tried to minimize it (e.g. Purox process.) Energetically, the organic compounds tend to concentrate product energy and careful regard must be paid to these chemicals when developing any system.

The 'oil' may be approximated chemically as $C_4H_9O_3$, although it contains a wide variety of oxygenated compounds from low to very high carbon numbers corresponding to tars or pitch. It is claimed to be compatible but immiscible with No. 6 fuel oil; it tends to polymerize and thicken to almost a solid state with time; and because of the oxygenated compounds it is very corrosive and requires special handling equipment. As far as is known, no work has been carried out on neutralizing or reacting this material to lessen the undesirable properties and produce a higher grade material. No other properties are known other than its fuel grade. When not required it is recycled to the reactor for total degradation (e.g. Purox process).

The aqueous phase consists of water and dissolved organic compounds. It is claimed to be biodegradable although the organics concentration can exceed 5% and includes some fairly intractable substances. As the water content of the feed material can be 30% by weight, the quantities of water for disposal are significant and must not be neglected.

The solid residue contains ash and other unreactive materials such as metal and glass. At relatively low reaction temperatures of below 900 °C, as occurs in most pyrolysis systems, there is considerable solid carbon present as char. A proportion of the energy content of the refuse can be lost in this way, unless a separate reaction system is installed. With air gasification, the carbon is oxidized leaving a relatively inert solid ash and all the energy containing products are vaporized. The higher temperatures produced in oxygen gasification cause the residue to melt to a fluid slag which can be run off.

The primary products from thermal processing are thus heat from incineration, and gas, 'oil' and possibly char from gasification/pyrolysis. All may be used as such or converted into higher grades of energy such as electrical, mechanical or

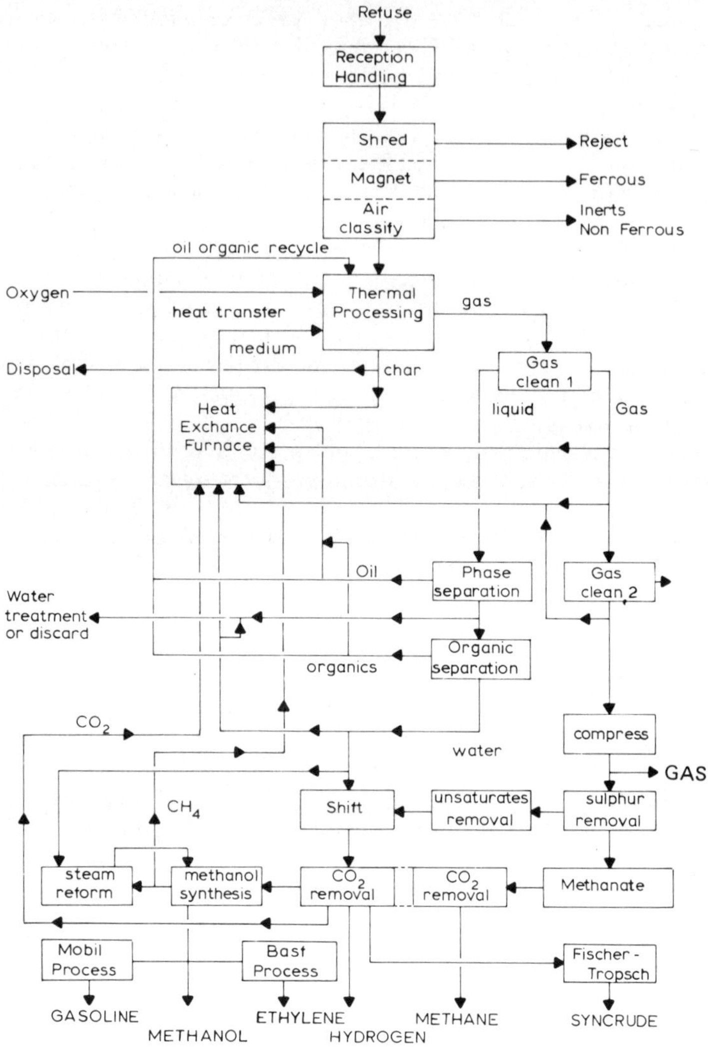

Fig. 12.2 A total processing system for turning refuse (or similar waste material) into fuel. (Reproduced by permission, from A.V. Bridgwater, *Resource Recovery and Conservation*, **5** (1980) 111.)

chemical. The total chemical system of converting refuse to high value chemical fuels or raw materials is shown in Fig. 12.2, which includes all possible interactions of side streams, by-products and recycle.

INCINERATION

The various methods for energy recovery from incineration are shown in Fig. 12.3. Heat recovery is the most widely practised (Fig. 12.3a) but is not without problems, as the following example shows. The City of Coventry, before incorporation into the County of West Midlands, UK, built an incinerator for the city's refuse with a

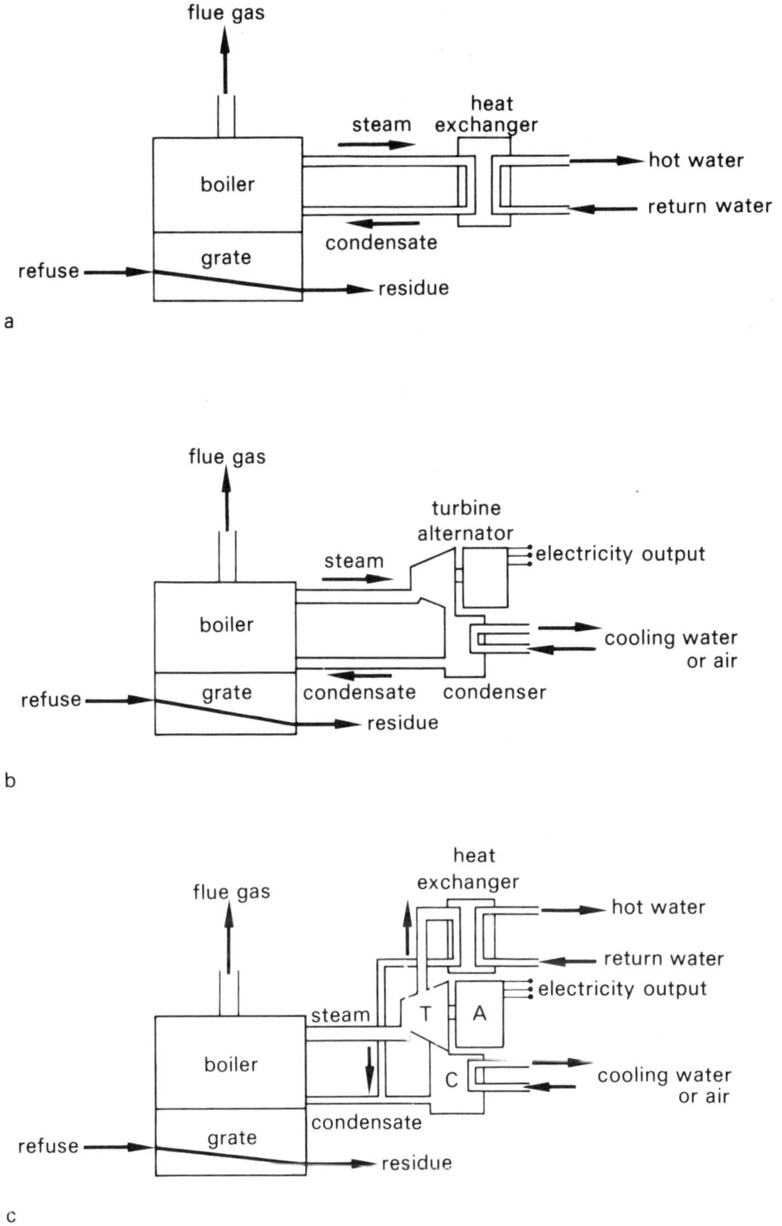

Fig. 12.3 Alternative methods of heat recovery: (a) district heating, (b) power generation, (c) combined power generation and district heating. (Reproduced from A. V. Bridgwater, *Waste Recycling and Pollution Control Handbook*, George Godwin/Van Nostrand Reinhold, by permission.)

contract to recover heat for the local Talbot (formerly Chrysler) car factory. An early problem was insufficient refuse, which was solved by 'importing' refuse from outside the city. There were serious corrosion problems, caused in part by return water at too high a temperature and in part by the nature of the materials being incinerated. More recently, the corrosion problem is understood to have become so serious as to warrant total replacement of most of the boilers at a cost greater than the total original cost. The viability of this particular installation cannot be considered satisfactory.

Generally, problems of 100% refuse incineration are related to corrosion from chloride attack and slagging, which is exacerbated by chloride mobilization of the slag, accelerating corrosion and reducing efficiency. Other problems relate to particulate carry-over, requiring more sophisticated flue gas clean-up systems, and erosion from abrasive particulates.

As up to 80% of the heat generated by incineration is recoverable and energy costs continue to escalate above the general inflation rate, heat and power recovery will become more viable. Electricity generation (Fig. 12.3b) is also becoming more popular, particularly the higher efficiency combined heat and power option (Fig. 12.3c) — the heat as hot water is suitable for district heating on a large scale, or industrial space heating on a smaller scale. Medium-to high-pressure steam may be employed for processing, higher-grade heating and/or electricity generation on a large-scale municipal or smaller-scale industrial level if arisings are sufficient to take advantage of the economics of scale.

Various options are possible, but with the overall limitation that disposal must be a continuous operation irrespective of seasonal changes in energy demand. More advanced proposals include heating and/or cooling for integrated air conditioning with package boilers on standby to cope with peak demands and a condenser to handle any surplus steam. Such a process is complex and costly, but shows what is required of a facility to be acceptable to the users.

In addition to incineration of refuse for both effective disposal and primary thermal energy production, a number of other applications have been developed.

In 1975, IMI Birmingham installed a supplementary refuse firing system in the company's coal-fired chain grate boiler. Raw refuse is shredded to − 75 mm in a vertical shaft shredder, followed by magnetic metal removal with an in-line overband magnetic separator. The shredded demetalled refuse is loaded by screw conveyor into one of two containers, which also compacts the refuse to about double the density. These containers transport the waste to the power plant about one mile away (1.5 km) and also provide a buffer stock to avoid the need for storage facilities in the unsuitable environment of the power plant. The refuse is transported on a variable speed, inclined slat conveyor to two burners situated in the rear wall of the burner over the coal fire already on the chain grate (Fig. 12.4) to give suspension firing with maximum retention time. Refuse and coal are fired in approximately equal quantities with a total capacity of 60 000 tonnes refuse/year.

Minor problems inevitably arose concerning particulates emission, chloride corrosion (which appeared to be significantly improved by co-firing with coal), water content which should not exceed 30%, and blanketing if excessive fines were added. The plant has successfully operated at a significantly higher level than 50% refuse and development work is continuing. The current feed material is industrial rather than domestic refuse, which reduces the fines problem. The economics appear very attractive with the raw refuse having a notional value of £7/tonne (1978).

IMI have cooperated on similar processes elsewhere both in the private and public sector. A current project is at Humberside, where a trommel screen has been installed after shredding to remove the − 25 mm 'fines', which also effectively removes a significant proportion of the putrescibles and hence reduces water content.

The second application is also a supplementary firing of refuse as a solid fuel substitute in cement manufacture. The process is very similar conceptually to that

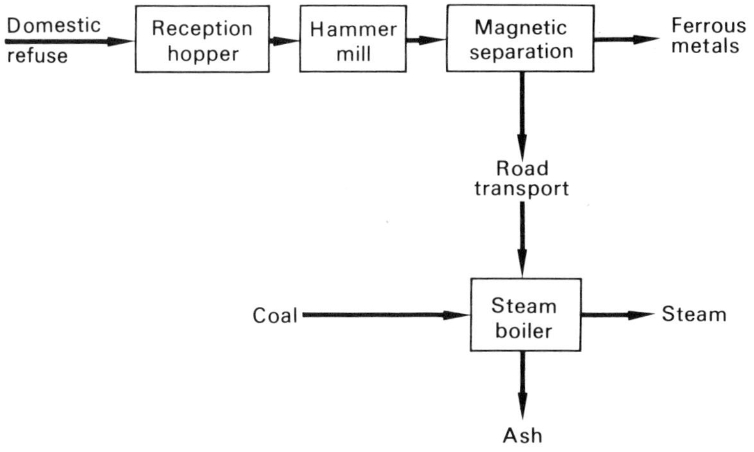

Fig. 12.4 IMI Witton refuse burning scheme (West Midlands County Council). (Reproduced from A. V. Bridgwater and C. J. Mumford, *Waste Recycling and Pollution Control Handbook*, George Godwin/Van Nostrand Reinhold, by permission.)

described above with shredding to −50 mm, screening, secondary shredding on the oversize, covered storage with front end loading to provide 24 h/day and 7 day/week operation, magnetic separation, and pneumatic conveyance to the fuel feed end of the rotary kiln (Fig. 12.5). Two plants are currently in operation at Shoreham and Westbury, and the process is still being developed. The economics are not fully evaluated but are understood to be interesting. Of key importance is the product quality which is understood to be affected but still well inside specification.

Many small boiler systems are now being installed to utilize locally arizing wastes such as wood and packaging materials at rates of 10 – 100 tonne/week to raise process steam and/or provide space heating requirements. Typical payback times are 2 – 4 years.

It is interesting and important to note that all the examples quoted provide a total service of raw refuse to useful end product without the need to market intermediates or products. It will be seen later how this differs in the gasification/pyrolysis developments.

GASIFICATION AND PYROLYSIS

As explained above, gasification employs a limited amount of air (air gasification) or oxygen (oxygen gasification) to burn part of the organic material, thereby releasing sufficient heat to thermally degrade (or pyrolyse) the remainder of the material. Pyrolysis can be considered as thermal degradation by indirect heating; for example by external heating or by contact with a heat transfer medium such as hot inert gas or molten salt. In all cases the following products are formed:

gas	carbon monoxide
	carbon dioxide
	hydrogen
	methane
	higher hydrocarbons
	nitrogen (if air or flue gases are added)
	contaminants such as chlorine and sulphur in some form

aqueous liquid	water dissolved organic compounds
non-aqueous liquid	tars and higher hydrocarbons
solid	char ash other non-combustible impurities such as glass or metal

It is usual to separate out inert materials, such as metals, glass, dust and stones, prior to thermal processing to avoid product contamination and increase throughput. The remaining organic fraction, consisting mainly of paper and board,

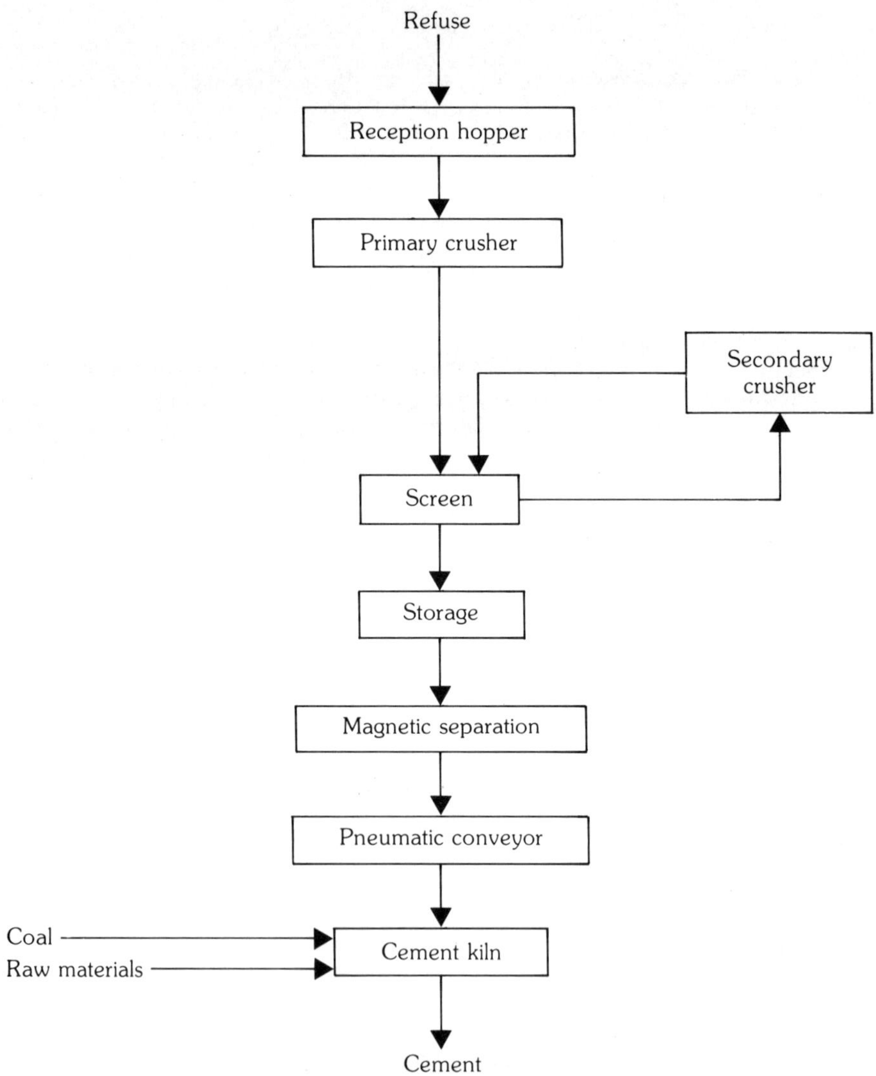

Fig. 12.5 Refuse-fired cement manufacture.

plastics, putrescibles and rag, is effectively RDF and may conveniently be approximated to cellulose in chemical composition, since cellulose is the principal ingredient of paper and putrescibles which together form the bulk of the organic fraction.

Chemistry

At high temperature, the cellulosic organic materials break down into four main products: fuel gas, pyrolytic oil, aqueous condensate and a carbonaceous solid residue. The proportion and composition of the products are affected by pyrolysis temperature, method of heating, rate of heating, contact time, recirculation, secondary processing, drying, type of equipment, catalysts, pressure and chemical additions such as oxygen, air, water, hydrogen, carbon monoxide or carbon dioxide. Information on the effects of these variables is still being gathered, and a surprisingly small amount of coherent information is available, although pyrolysis has been practised since before 1800 to produce gas from coal.

Generally, reaction temperature is one of the most important variables in controlling the proportion of the four products. Pyrolysis begins at around 250 °C and becomes significant above 300 °C. As temperature increases, greater proportions of gas are formed and less liquid distillate and solid char.

Some typical figures of product composition are shown in Fig. 12.6, which assume that no products are recycled internally or externally. The effect of rate of heating is also shown, high heating rates favouring oil production.

The basic chemical reactions may be represented by the degradation of cellulose in the following approximations:

For gasification from 1100 to 1500 °C

	gas		water		oil/tar		char
$6C_6H_{10}O_5 \rightarrow$	$11CO + 10CO_2 + 6H_2$	$+$	H_2O	$+$	$\tfrac{1}{2}C_4H_9O_3$		

For pyrolysis from 500 to 900 °C

	gas		water		oil/tar		char
$6C_6H_{10}O_5 \rightarrow$	$7CO + 5CO_2 + 3H_2 + CH_4$	$+$	$6H_2O$	$+$	$3C_4H_9O_3$	$+$	$8C$

Application

Considerable effort has been put into thermal processing of wastes both as an economical and satisfactory means of disposal, as well as production of a higher grade fuel than the waste itself. Over 150 areas of work have been identified around the world, ranging from bench-scale experimentation to substantial demonstration-size plant capable of handling hundreds of tonnes per day. Although a few processes are available commercially, only one, to date, is understood to have been installed and operated successfully.

Most processes are derived from analogous coal processing systems, with

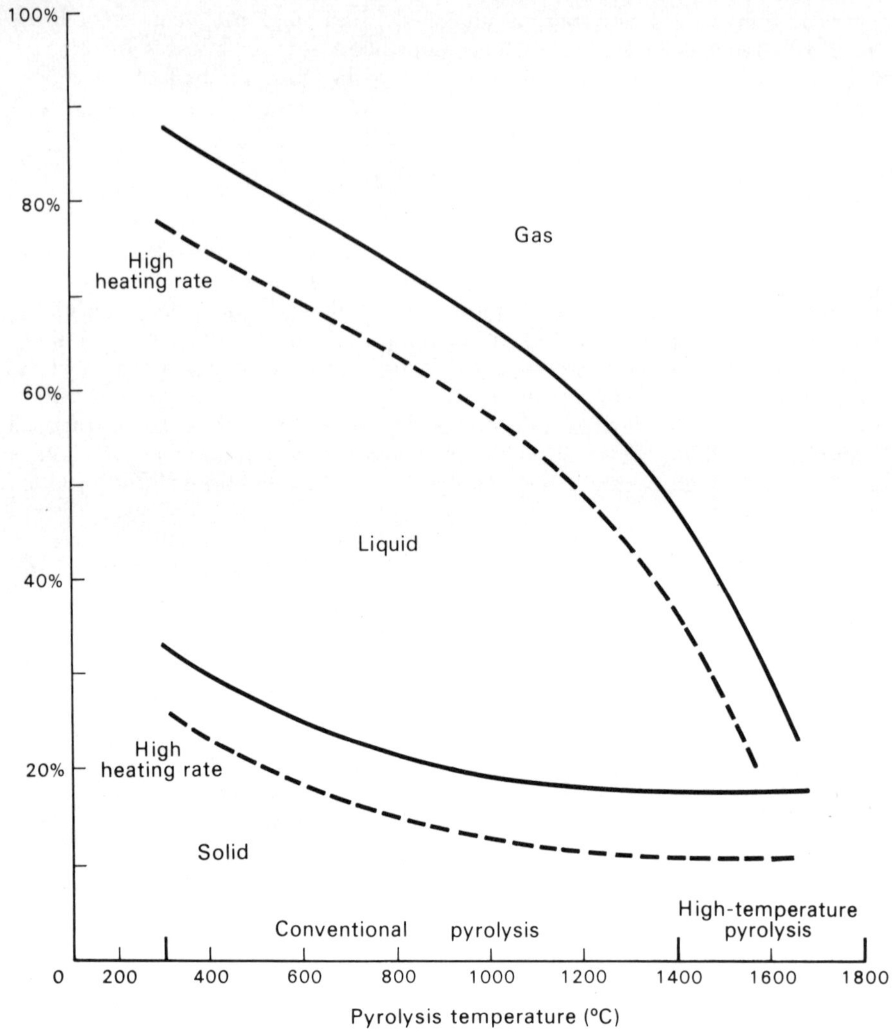

Fig. 12.6 Effect of temperature on proportions of products. (Reproduced from A. V. Bridgwater and C. J. Mumford, *Waste Recycling and Pollution Control Handbook*, George Godwin/Van Nostrand Reinhold, by permission.)

modifications to allow for the different physical and chemical characteristics of the feed. It is interesting, however, that not all of the processing variations tried on coal have yet been applied to waste, nor has the concept of an integrated coal to fuel been applied to waste to fuel. There are some interesting possibilities here which are explained later.

The current situation is that limited research is currently being carried out in a number of centres in North America and Europe with some work also in Japan. Much of the early work was poorly reported and there are many inconsistencies and even contradictions in published results. The primary objective of most of that work was to use pyrolysis and gasification as disposal techniques. Now that such an approach is discredited on economic and technical grounds, thermal processing is seen as a longer-term solution for recycling and recovery. Current

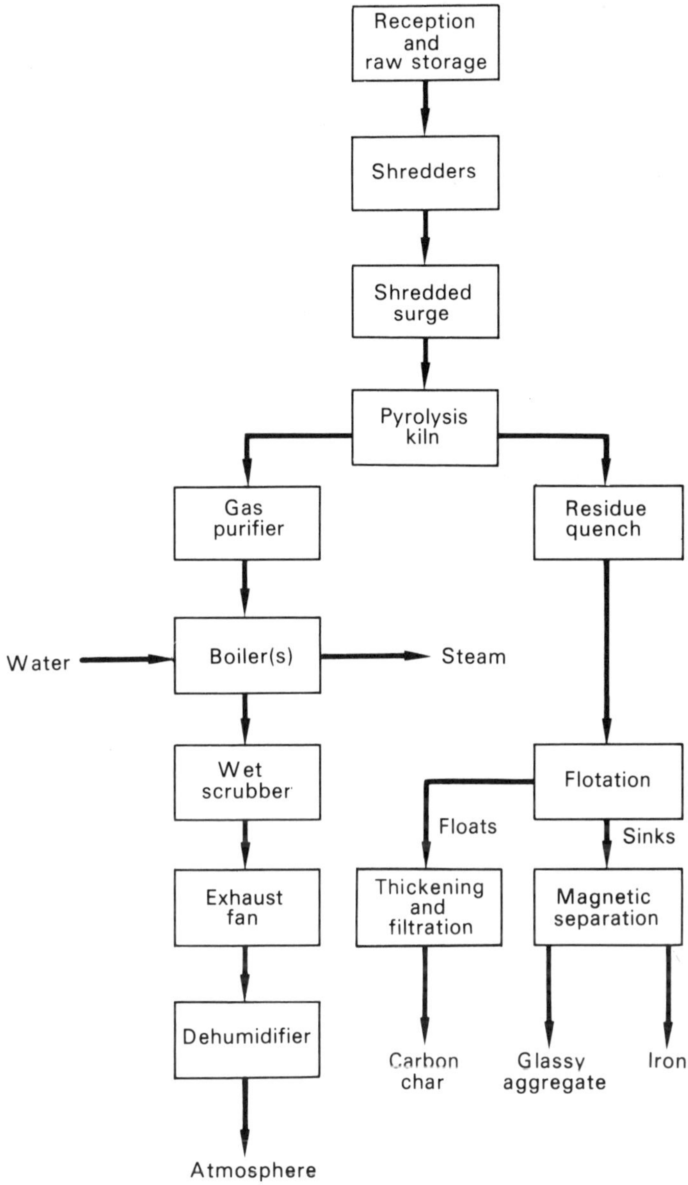

Fig. 12.7 Landgard (Monsanto-Envirochem) resource recovery and pyrolysis system. (Reproduced from A. V. Bridgwater and C. J. Mumford, *Waste Recycling and Pollution Control Handbook*, George Godwin/Van Nostrand Reinhold, by permission.)

work is thus trying to consolidate information and obtain a more thorough understanding of the principles involved.

There are several systems at an advanced stage of development and available commercially that justify description and comment.

The only process commercially installed is the Landgard process which produces gas and char only in an oil-fired horizontal rotary kiln at a maximum

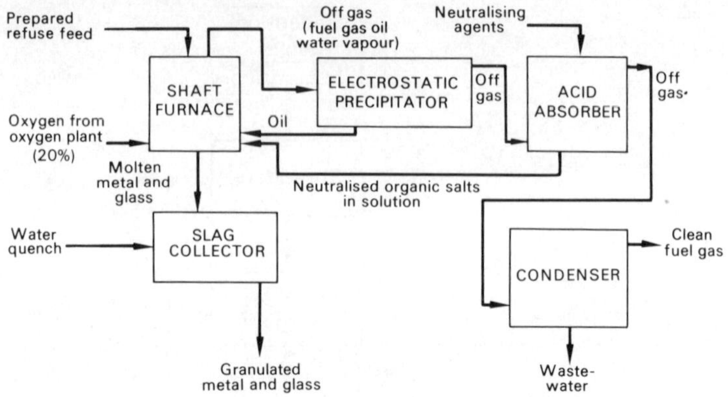

Fig. 12.8 Purox (Union Carbide) pyrolysis process. (Reproduced from A. V. Bridgwater and C. J. Mumford, *Waste Recycling and Pollution Control Handbook*, George Godwin/Van Nostrand Reinhold, by permission.)

temperature of 980 °C (Fig. 12.9). The gas is normally burnt to raise steam, although it could be otherwise used. The only pretreatment is shredding; some post-pyrolysis treatment may be practised to recovery ferrous metals, carbon char and glassy aggregate. Oil production is minimal by design, in that any oil formed is thermally decomposed to gas. This is one of the few refuse pyrolysis systems to be installed in a 'live' situation, in Baltimore, at about half design capacity of 500 tonne/day. To meet air pollution requirements, additional gas-cleaning equipment (electrostatic precipitator) has been installed. The process is being operated as two stage incinerator to raise steam, but the gas could be exported.

The Purox (Union Carbide) process is one of the most widely publicized (Fig. 12.8). This employs oxygen gasification in a vertical shaft furnace, using pure oxygen rather than air to avoid dilution of the pyrolysis gas with nitrogen. The process differs from that previously described in that a slag is formed which is run off to a water quench. The maximum temperature is about 1500 °C and only two products are formed — gas and solid. The small amount of oil formed is returned to the reactor. The gas represents over 80% of the fuel value of the incoming refuse and, even after deducting in-plant requirements, about 70% is available as surplus. Resource recovery may be carried out as a pretreatment operation or on the solid product to recover metals and glass. Although this is a well developed and technically attractive process, the very high initial costs are precluding commercial development. A plant of capacity around 1000 tonne/day would cost nearly £40 million (1980). Recent studies suggest that the UK is not yet ready for a process of this size or cost.

The Andco-Torrax (Carborundum) process (Fig. 12.9) is a high-temperature slagging system (like Purox) and uses all the gas, either in-plant or to export energy (like Landgard), rather than exporting surplus gas. Processing is based on air gasification using preheated air, giving a relatively low-grade fuel gas product. This avoids the need for expensive oxygen generation, but at the cost of producing a less valuable product which has to be used immediately to avoid loss of sensible heat. One plant is already in operation, in Luxembourg, operating at just over half design capacity of 180 tonne/day, with further plants

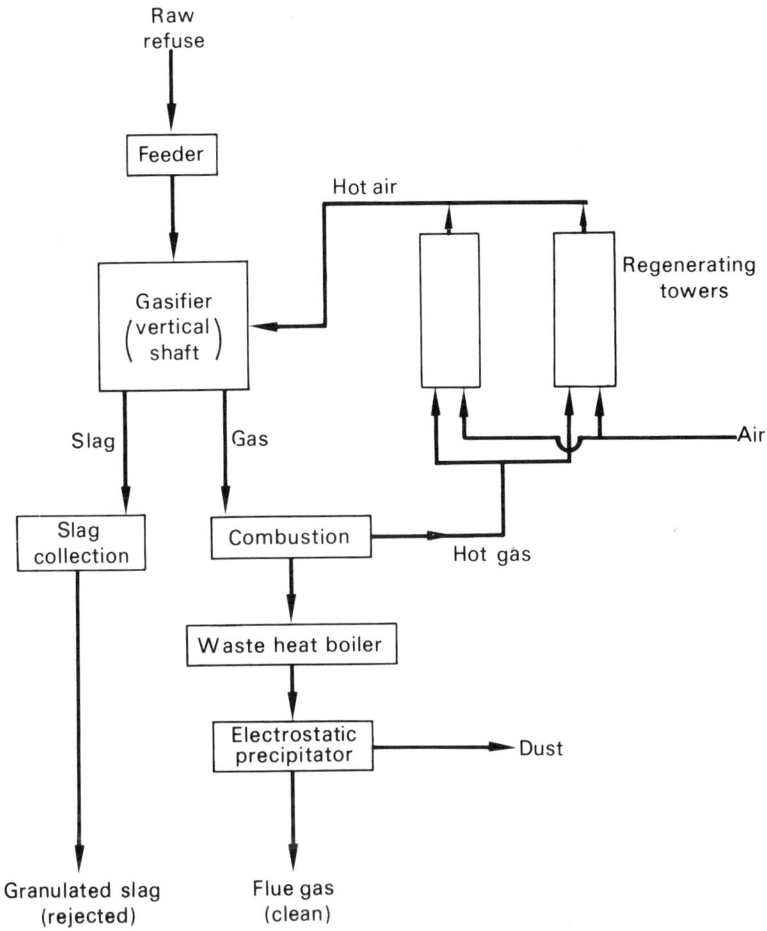

Fig. 12.9 Andco-Torrax (Carborundum) pyrolysis system. (Reproduced from A. V. Bridgwater and C. J. Mumford, *Waste Recycling and Pollution Control Handbook*, George Godwin/Van Nostrand Reinhold, by permission.)

at Grasse and Creteil, France, and Frankfurt, W. Germany. As this is essentially a two-stage incineration process, it is difficult to understand its adoption in Europe where air pollution standards tend to be less severe than in the USA.

Warren Spring Laboratory produced the Cross-Flow reactor system (Fig. 12.10), which is currently being developed and marketed by Foster Wheeler Power Products for tyre pyrolysis as well as refuse or biomass pyrolysis. This system was developed after research on laboratory scale retort pyrolysis and induction heating with circulating steel balls. To obtain better heat transfer to the refuse (one of the more serious problems), some of the gas produced is recycled, with preheating, to provide a heat transfer medium. Char is removed from the pyrolysis reactor and oil and aqueous fractions are condensed out. While some of the gas is used for recycling and predrying the feed material, at least 50% should be available surplus to in-plant requirements. Resource recovery could be practised before or after pyrolysis.

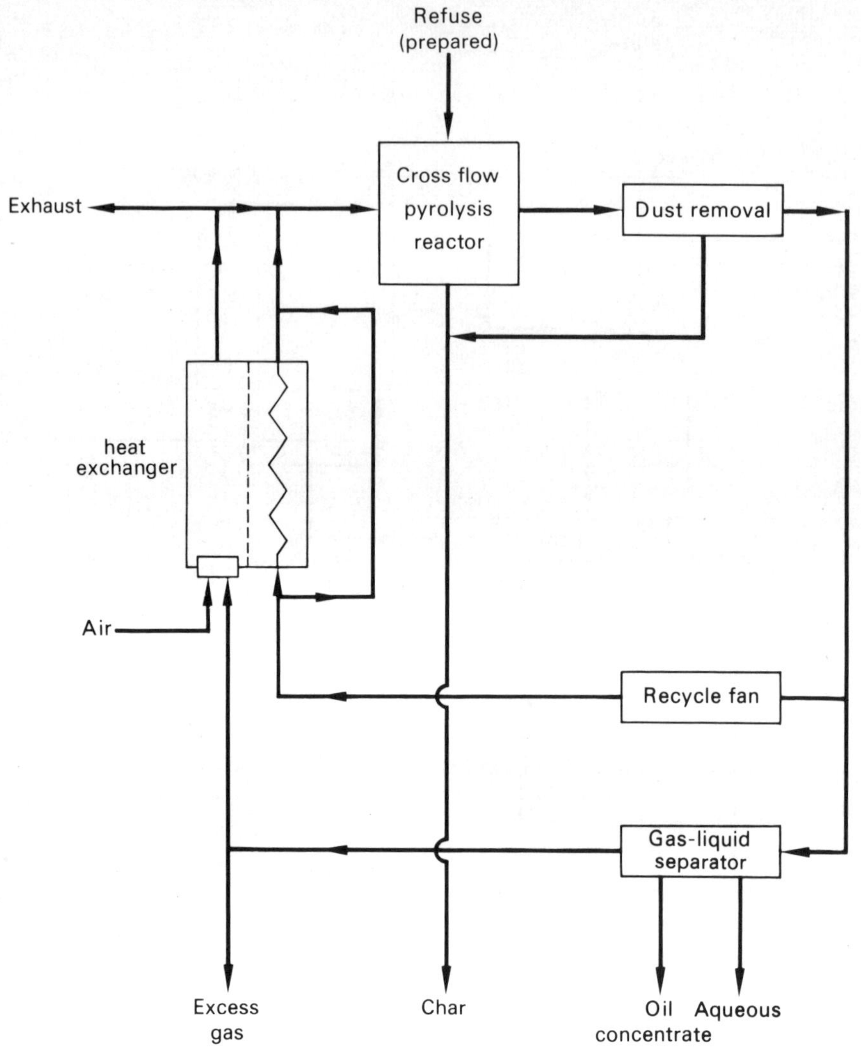

Fig. 1210 Pyrolysis by hot product gas recycling — cross-flow system (Warren Spring Laboratory/Foster Wheeler Power Products). (Reproduced from A. V. Bridgwater and C. J. Mumford, *Waste Recycling and Pollution Control Handbook*, George Godwin/Van Nostrand Reinhold, by permission.)

PRODUCTION OF FUELS AND CHEMICALS

It has already been explained that early work regarded thermal processing as an attractive disposal route. The concept of utlizing the products either directly or after conversion to higher grade products was not widely appreciated. As development proceeded, first the primary products were utilized, usually as a fuel gas, although the Occidental (Garrett) process produced a fuel oil compatible with No. 6 fuel oil, and later the advantages of producing high grade liquid or gaseous fuels/chemical intermediates were investigated. The Purox system, for example, featured in many paper studies for converting wood or refuse into methanol or ammonia.

The gas from thermal processes is a synthesis gas which many years ago formed one of the main chemical production routes. After cleaning and purifying the gas, the composition may be chemically adjusted to give gas of the correct proportions for making methanol, methane, hydrogen or synthetic crude oil/mixed hydrocarbons. Recent developments convert methanol into gasoline or ethylene.

The total processing system of turning refuse (or any similar waste material) into any of the above fuels or chemicals was shown in Fig. 12.2. All the significant steps, from feedstock to end product, are shown in a feasible sequence of operations. There are many options available, all of which are shown, for utilization and/or treatment of any primary products of thermal processing. For simplicity, multiple stream operation such as char gasification or production of methanol and methane is omitted, although feasible and possibly economically attractive.

Although it is convenient to regard the total system as a front-end thermal process with a back-end conversion process, it is not really quite so simple. There are some real and apparently attractive possibilities for recycling products and by-products to the thermal process. Chemical, energetic and economic optimization is necessary.

For preliminary design and costing there are two simple rules:

1. Keep the process as simple as possible by reducing the number of steps to a minimum; a well known strategy for preliminary design and crude economic optimization.
2. Recycle as much as possible and/or keep losses to a minimum as far as feasible and consistent with keeping processes simple — a crude energy and chemical optimization.

Some of the technical and economic design parameters in converting synthesis gas to useful higher value products are exemplified by reference to methanol, methane hydrogen, and liquid hydrocarbons.

METHANOL PRODUCTION

As an example of the technical and economic viability of converting organic waste to a valuable chemical feedstock or fuel, the process in Fig. 12.11 is described and costed.

Refuse is first shredded and the thermally degradable components separated. This material is then oxygen gasified at high temperature to give complete conversion to gas. Oxygen gasification is specified as it avoids 'oil' production, which is difficult to manage, and a full scale process is commercially available (Purox). After washing to remove tars, organics, dust and water, the gas is further cleaned in an electrostatic separator. This clean raw synthesis gas is now available for methanol production in a conventional low pressure process.

There are many assumptions employed in deriving the front-end sequence, of which the most significant are minimum pretreatment of refuse, maximum production of gas, recycling of organic liquid products, minimum water treatment requirements, and only two stages of gas cleaning. This conceptual process and the assumptions employed have been derived from careful study of many of the proposed and developed processes.

Methanol production from the resultant synthesis gas is well understood but

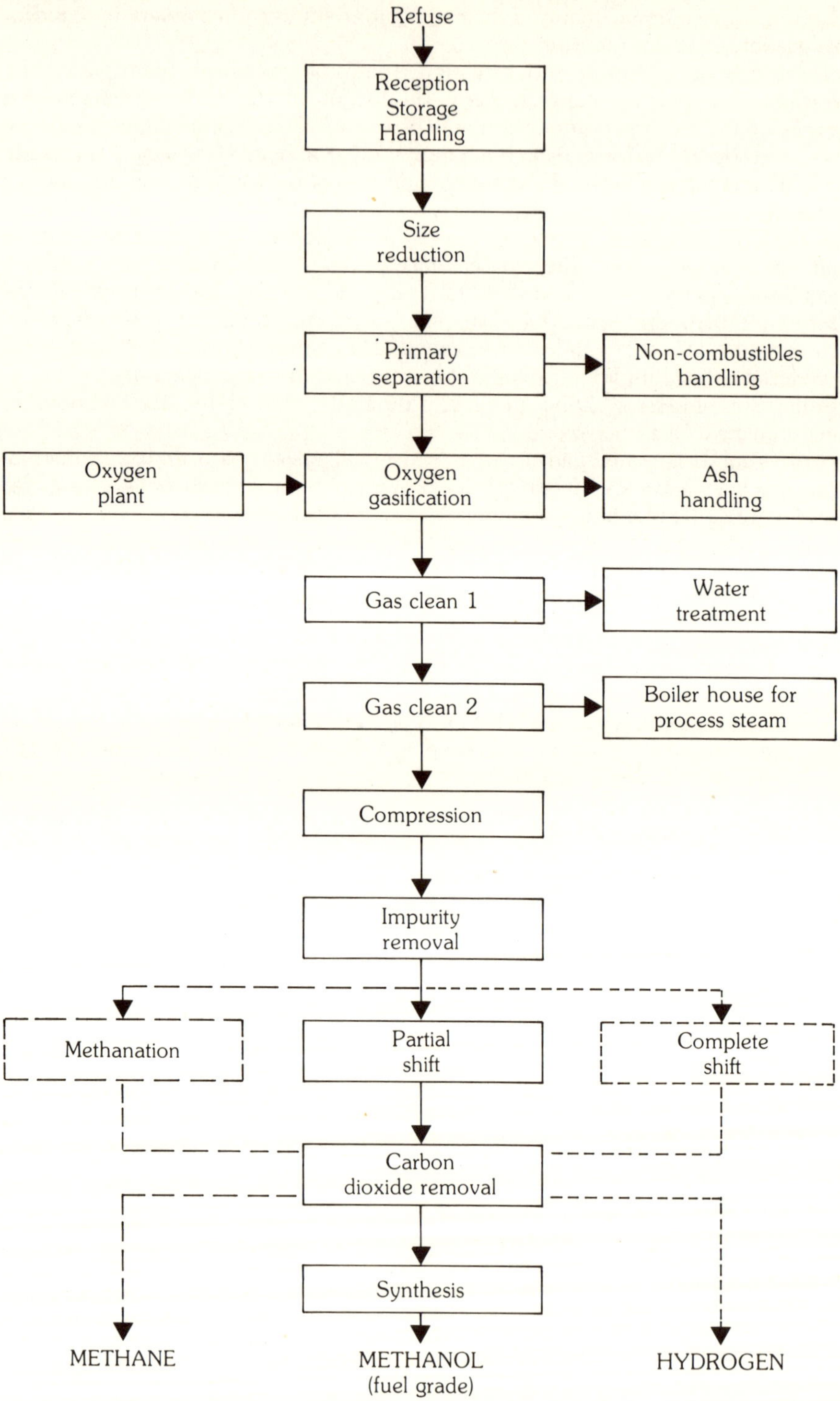

Fig. 12.11 Conceptual waste-to-methanol process.

has a sequence of operations significantly dependent on the gas composition: for example, the carbon monoxide to hydrogen ratio ($CO:H_2$) is stoichiometrically important — the $CO:H_2$ ratio may be adjusted by the shift reaction, but this produces more carbon dioxide and reduces the carbon efficiency; the carbon dioxide content is important — a minimum level of around 3% is necessary but excess requires removal; methane need not be removed until after synthesis, but then requires purging to be reformed by synthesis gas, used as on-plant fuel or exported. Without a sound appreciation of the relationship between gas composition and thermal processing parameters, it is difficult to specify the optimal sequence of operations overall for methanol synthesis.

The minimum process requirement for methanol synthesis is a gas of low carbon dioxide level and correct ratio of carbon monoxide to hydrogen: only compression, sulphur and unsaturated hydrocarbon removal, and methanol synthesis are required. This is unlikely to be achieved in any reaction system, and the most likely arrangement will also require a shift reactor to give the correct $CO:H_2$ ratio and carbon dioxide removal before the methanol synthesis. The most expensive alternative will include a methane reformer to give more synthesis gas. There is a considerable range of costs and yields associated with these options. A fuel-grade methanol of around 98% purity is produced. Higher grades are produced by one or two stages of distillation.

Cost Premises

Detailed techno-economic evaluations have produced the following yields and production costs for methanol at two scales of production and with two feedstock/raw material costs. One basis assigns no cost or credit to the refuse, and the other assumes a pretreated refuse cost of £15/tonne of thermally degradable fraction and feedstock, i.e. RDF. All process energy requirements are assumed provided in-house, and suitable allowances have been included in the evaluatory studies for this provision. The production costs assume 100% debt or loan capital finance at 10% over 10 years with no allowance for any return.

Production and cost data (mid-1980 basis):

Refuse feed		
wet	1000 tonne/day	
dry	700 tonne/day	
Thermally degradable fraction	450 tonne/day	(at 65% content)
Methanol produced	225 tonne/day	(at 50% weight conversion of degradable fraction to methanol)
Capital cost	£27.8 million	(battery limits basis, 1980, UK)
Cost per tonne		
methanol at zero raw material cost	£129/tonne	
at feedstock cost of £15/tonne	£157/tonne	

At 2000 tonne/day, which can probably be regarded as the maximum possible sized plant based on refuse feed, the data on the same basis become:

Refuse feed	
wet	2000 tonne/day
dry	1400 tonne/day
Thermally degradable fraction	900 tonne/day
Methanol produced	450 tonne/day
Capital cost	£38.9 million
Cost of methanol	
at zero feedstock cost	£87/tonne
at £15/tonne feedstock cost	£130/tonne

The current manufacturing cost of methanol is about £125 for a newly constructed plant.

These figures are for an established technology without any optimization beyond that described above. New technologies being developed and careful optimization could ultimately improve weight conversion by 10% or more, giving a methanol cost at 1000 tonne/day raw refuse feed of about £120/tonne, and about £80/tonne at 2000 tonne/day raw refuse feed.

The current real inflation of energy costs, evidenced by the UK government's declaration of natural gas prices rising at 10% above background inflation for the next three years, mean that orthodox methanol production from natural gas will escalate at least 5% above background inflation as raw material costs are so significant. The cost of methanol from a new facility in three years time at 1980 prices will thus be around £160/tonne.

Extension of such sensitivity studies have some value but at this point the economics should perhaps only be regarded as interesting. Other significant factors are the rapidly escalating cost of conventional refuse disposal and contrarily the reluctance of authorities to be involved in new technology and at such a high capital cost. There is also the problem that a chemical commodity is produced for which there is already overcapacity, and without a clear policy on such products implementation is unlikely.

METHANE PRODUCTION

An analogous process to methanol production is shown on the left of Fig. 12.11. The thermal processing front end is identical to that for methanol production and a variety of methanation processes are available although none is yet commercially proven. The methane product may be employed as a fuel — SNG — or as a chemical feedstock.

Production and cost data (mid-1980 basis):

Refuse feed		
wet	1000 tonne/day	2000 tonne/day
dry	700 tonne/day	1400 tonne/day
Thermally degradable fraction	450 tonne/day	900 tonne/day
Methane produced	93 tonne/day	186 tonne/day
Capital cost	£25.4 million	£39.8 million
Cost of methane	£475/tonne	£364/tonne
at zero feedstock cost	£315/tonne	£240/tonne
at £15/tonne feedstock	£439/tonne	£364/tonne

The sensitivity analysis and discussion are very similar as for methanol. The conversion efficiency is very low because of the poor carbon utilization, which is a consequence of unsuitable gas composition. This illustrates well the need for thermal process optimization for a specific end product. For methane production, for example, steam gasification gives much better yields.

HYDROGEN PRODUCTION

This is also shown in Fig. 12.11 to the right. All the carbon monoxide is shifted by reaction with steam, and the resultant carbon dioxide removed. Although such a process cannot be considered as seriously as methanol and methane, for fuel production, it demonstrates the versatility of thermal processing.

Production and cost data (mid-1980 basis):

Refuse feed		
wet	1000 tonne/day	2000 tonne/day
dry	700 tonne/day	1400 tonne/day
Thermally degradable fraction	450 tonne/day	900 tonne/day
Hydrogen produced	48 tonne/day	96 tonne/day
Capital cost	£28.6 million	£44.9 million
Cost of hydrogen at zero	£680/tonne	£520/tonne
feedstock cost	£0.061/N m^3	£0.046/N m^3

SYNTHETIC CRUDE OIL PRODUCTION

An analogous process can be conceived for syncrude production via the Fischer-Tropsch synthesis. The thermal process front end would be similar with the synthesis gas going directly to the reactors. Although well developed in South Africa, few details are available. The following data are thus a little more speculative than previous estimates for methanol, methane and hydrogen (mid-1980 basis):

Production and cost data:

Refuse feed	
wet	1000 tonne/day
dry	700 tonne/day
Thermally degradable fraction	450 tonne/day
Syncrude produced (as liquid hydrocarbons)	33 tonne/day (7% weight conversion)
Total hydrocarbons produced	115 tonne/day (29% weight conversion)
Capital cost	£42 million
Syncrude cost as liquid hydrocarbons at zero feedstock cost	£1473/tonne (£1151/tonne at 2000 tonne/day raw refuse feed)

The product would need to be refined as for crude oil, although it is believed to be possible to run the process partially selective to required products. This is not a serious alternative.

GASOLINE

Gasoline (petrol) may be produced selectively at very high efficiency from methanol or directly from synthesis gas employing new technology developed by Mobil. Based on a methanol intermediate, the capital, conversion and total product costs for gasoline are:

Production and cost data

Refuse feed (wet)	1000 tonne/day	2000 tonne/day
Methanol produced	225 tonne/day	450 tonne/day
Methanol Cost		
at zero feedstock cost	£129/tonne	£87/tonne
at feedstock cost of £15/tonne	£157/tonne	£130/tonne
Gasoline produced (38% weight conversion)	86 tonne/day	172 tonne/day
Incremental capital cost of gasoline plant 'add-on'	£4.5 million	£7.2 million
Incremental production cost of gasoline, per tonne gasoline	£58/tonne	£45.4/tonne
Total production cost of gasoline	£398/tonne	£274/tonne
at zero feedstock cost		
at feedstock cost of £15/tonne	£471/tonne	£388/tonne

To produce gasoline competitively with current natural gasoline at the higher throughput plant requires a methanol production cost of £60/tonne.

FUEL GAS

A much more attractive short-term possibility for smaller waste arisings is low-grade fuel gas production. Some systems are available with others under development, and the fuel gas can be used locally for steam, power or electricity generation. The economics are dependant on a low-cost reliable gasification system being available, but this is not seen as a serious problem.

CONCLUSION

Incineration with heat and/or power recovery is becoming more widely acceptable, and with real energy costs continuing to increase will become a more viable option. Gasification and pyrolysis, however, are more complex and costly. Their significant advantage is the ability to considerably upgrade the energy content of waste into a premium value product — either a medium heating value gas for local use, or a liquid fuel/chemical feedstock. Implementation is unlikely to occur for some considerable time although, rightly, research continues.

ACKNOWLEDGEMENT

Much of the content of this chapter is taken from *Waste Recycling and Pollution Control Handbook* by A.V. Bridgwater and C.J. Mumford, published by George Godwin in the UK, and Van Nostrand Reinhold in the USA.

BIBLIOGRAPHY

1. ADER, G., BRIDGWATER, A.V. and B.W., *Conversion of Biomass to Fuels by Thermal Processes*, report to ETSU, Department of Energy, Phase 1 Review and Preliminary Assessment, 1978.
2. BRIDGWATER, A.V., HATT, B. and ADER, G., 'Chemical Fuel and Raw Material Production by Thermal Processing of Refuse — Technology and Economics', *Proceedings of the International Recycling Congress*, Berlin, 1979.
3. HATT, B.W. BRIDGWATER, A.V. and ADER, G., 'Refuse to Fuels — an Appraisal of Thermal Processes.' *Proceedings of the International Recycling Congress*, Berlin, 1979.
4. BUEKENS, A.G. and SCHOETER, J.G., *European PTGL Technology*, presented before the Division of Environmental Chemistry, American Chemical Society, in Washington DC, 1979.
5. ADER, G., BRIDGWATER, A.V. and HATT, B.W., *Conversion of Biomass to Fuels by Thermal Processes*, Optimisation Studies for Conversion to Methanol, report to Energy Technology Support Unit, Department of Energy, April 1980.
6. ADER, G., BRIDGWATER, A.V. and HATT, B.W., *Conversion of Biomass to Fuels by Thermal Processes*, Optimisation Studies for Conversion to Methane, report to Energy Technology Support Unit, Department of Energy, July 1980.
7. ADER, G., BRIDGWATER, A.V. and HATT, B.W., *Conversion of Biomass to Fuels by Thermal Processes*, Optimisation Studies for Conversion to Liquid Hydrocarbons, report to Energy Technology Support Unit, Department of Energy, September 1980.

Dr. Tony Bridgwater is Senior Lecturer in the Department of Chemical Engineering at the University of Aston in Birmingham. After working for several years with a major oil company, he joined the staff of the university and has developed interests in a wide range of business aspects of the chemical and process industries and also resource conservation, recycling and environmental protection. He is currently founder editor of *Process Economics International*, after also being responsible for launching *Engineering and Process Economics*. He is also a member of the editorial board of *Resource Recycling and Conservation*. He has published over 70 papers in the above areas and recently his interest has been focused on the recovery of valuable forms of energy from organic waste including biomass. His major publications include *Waste Recycling and Pollution Control Handbook*, and *Chemical Waste and By-Product Recovery*, which is shortly to be published.

13 Measures Intended to Influence Quantity and Quality of the Packaging Stream in Household Waste

Karl Lidgren

PRICING OF REFUSE-COLLECTION SERVICES

For the collection of household waste, consumers in Sweden, by means of a refuse collection levy, pay compensation for the resources consumed in manpower, fuel, etc. (In many other countries there is no refuse collection levy, but this activity is paid for through some public budget and is thus paid indirectly via income tax.) No levy for the water, air and land pollutions that arise in connection with waste handling is charged, however. These types of effect on the environment from packaging systems are dealt with later.

Nowadays, refuse collection levies are based on a certain amount per unit of volume independent of the refuse's quality, although differing qualities can give rise to different handling costs. But, as in practice it is almost impossible to decide who throws away what kind of waste, there may be reason to investigate whether the refuse collection levy should not instead be placed directly on the product. Thus, unlike the current refuse collection levies, those of 'tomorrow' should be paid at the time of purchase and be dependent on both quantity and quality of the waste finally constituted by the packaging. A procedure of this kind presumes also, however, that there really are different waste handling costs for different types of packaging, and that it is known in advance how large these are. Another fact that also favours a reorganization of refuse collection charges is that many consumers, in multifamily houses, pay only an average levy for the house (or area), which in addition is most often difficult to 'realize' as it is included in the rent. In such a case, a consumer who doubles his waste volume does not undergo a doubling of the waste collection levy, but only the perhaps insignificant increase of the average refuse collection levy that affects everyone in the collective. A similar state of affairs also appears when the refuse collection costs are paid through some public budget. This applies to dustbins put out along roads and beaches, in parks, etc.

The economic control measure to be discussed here in more detail is the possibility of introducing a 'perfect' refuse collection levy for, among other things, beverage containers. 'Perfect' refuse collection levy here means a levy corresponding to the national economic costs for taking care of waste, and is considered to be paid by the consumer at the time of purchase and not, as today, when he disposes of waste.

The administrative measure to be investigated is the introduction of compulsory separation of certain beverage containers.

This chapter begins to discuss the possibility and desirability of altering the individual's cost for disposing of his refuse through the introduction of a 'perfect' waste collection levy, which should be included in the price the consumer pays when purchasing the article that is presumed to land later on, in the household waste.

Refuse collection levies are currently based on a certain amount per unit volume, independent of the quality of refuse. (Photograph by Bengt Olsson).

The 'perfect' waste collection levy should be included in the price the consumer pays when he buys goods. (Photograph by Bengt Olsson).

Even if the container finds its way to some kind of recycling treatment, there are variations in the disposal costs of different kinds of container and of containers of the same type (dependent on which waste treatment system is used).

When a container is sold, the price is naturally of importance, especially if a substitute is available. This implies that it is not likely that the container manufacturer will accept increased costs to make his container less expensive in the waste handling sector. A buyer of a consumer container is presumably not prepared to pay a higher price, for example, for one that is recoverable, even though from the national economic view this should be cheaper. The disposal cost, that is to say the waste collection levy if the container is thrown away, is in fact mainly based on the waste's volume and completely dependent on whether that volume is filled, for example, by recyclable waste. The individual's ability to be sure of decreasing his total container costs (the cost of purchase plus the cost at the time of disposal) through a lower refuse collection levy because of 'better' waste is thus excluded until this levy system is altered from being based solely on quantity to being based also on quality. To this must be added the problem of when the waste collection levy is paid by some collective.

SHOULD LEVIES FOR REFUSE COLLECTION BE CHANGED?

Every municipality in Sweden has a legal right to exact levies for the collection of, and responsibility for, household waste. The levy may be set so that it covers all costs. On the other hand, the municipality may not profit from these activities overall. In the basis for the levy, however, the municipality's payment to contractors may be included, although these must be calculated with a certain amount of profit in their work. In addition, the municipality is entitled to carry out cost equalization between those who pay for the service.

The waste producers (households) thus pay together, through refuse collection levies, the total financial municipal costs for taking care of household waste. It should be observed here that in these financial municipal costs no payment is included for environmental effects arising in connection with waste management. The total sum debited to the household is thus less than equivalent to the total national economic costs.

Where possible, the individual household is debited directly for the waste collection levy appropriate for that household alone. In various kinds of collective household, this procedure is often impossible, so in such cases the waste producer pays a refuse collection levy that is only an average for the collective. The above shows that the refuse collection levies debited

1. do not correspond to all national economic costs for taking care of the waste,
2. are not based on the waste's quality,
3. are not always debited in relation to the waste production of the levy's payer (in collective debiting).

This has the effect that no greater encouragement is given to either the producer or the consumer to use products that result in qualitatively better waste. Furthermore, the levy system does not create any greater encouragement for certain households (and this also affects the producers) to keep down the quantity of waste. Against this background the following question can therefore be asked:

> Are there reasons and possiblities for changing the waste collection levy system, so that some encouragement is created for the development and use of beverage containers that are less costly in the waste handling process compared with other beverage containers?

It appears obvious that if a reconstruction of the refuse collection levy system is to take place, this reconstruction should not be limited solely to covering certain beverage containers. Further, the situation is such that, when quality control is impossible to carry out in practice when the waste is thrown into some receptacle, especially if the container is common to a collective, the quality aspect must be dealt with prior to the waste being thrown away. This means that the alternative to the current waste collection levy system, based on a levy at disposal, is a levy system based on the levy being paid at purchase of the product that will later be disposed of in the waste. The levy should therefore be fixed so that both qualitative and quantitative aspects are considered. The question, however, is how realistic such a 'perfect' levy is and whether any profit for society would arise in a reconstruction of the levy system.

As can be seen in Table 13.1, household waste consists of 18% kitchen waste (e.g. the results of fish cleaning, apple peel, old food). It seems to be difficult to find any way to reduce this kind of waste. The same can also be said to apply to sections of the other items: paper 48% (e.g. books), plastics 5% (e.g. toys), textiles 2% (e.g. old stockings), rubber and leather 1% (e.g. old bicycle tubes), glass 9% (e.g. broken drinking glasses), iron and metals 5% (e.g. old toasters), and miscellaneous 12% (e.g. old car batteries). This signifies that for a large part of the household waste, perhaps over half, it is very difficult to find a basis for calculation in a reconstruction of the refuse collection levies so that the consumer pays these at the time of purchase. It was pointed out earlier that containers constitute only 25% by weight of household waste.

A further difficulty arises when the costs for dealing with household waste vary because of the method of processing and the size of the processing plants. Fig.

TABLE 13.1 Household Waste Composition

Average water content 20 – 25%.

	% by weight
Paper	48
Plastics	5
Textiles	2
Rubber, leather	1
Glass	9
Iron and other metals	5
Kitchen waste	18
Miscellaneous	12
Total	100

13.1 shows the wide spread that occurs in regard to handling costs. In practice it is of course impossible to consider this with a levy system based on a levy at purchase. Fig. 13.1 also shows that transport and collection costs are responsible for approximately half the total waste handling costs. Obviously there is a connection between transport and collection costs and the quantity of waste, but this relation must not be overvalued, as the collection frequency must be kept to a certain level for hygienic reasons.

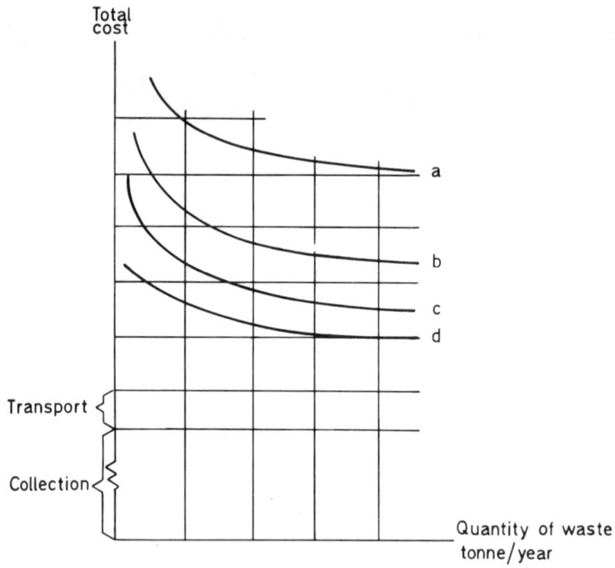

Fig. 13.1 The approximate principle cost picture for different methods of waste handling: (a) burning, (b) composting, (c) stacking with prior treatment, (d) stacking without prior treatment.

This level also tends to be independent of the selection of beverage containers, and of other components for which it is possible to debit a refuse collection levy in advance. This applies, however, mainly to individual households, as in certain rental areas it is possible that a decreased quantity of waste could lead to a collection run being deleted.

In the case of the handling alternatives of stacking and burning, only glass bottles can cause greater problems than the average waste. This occurs in the actual waste collection, where injuries (cuts) can occur to the personnel who collect certain types of refuse container (sacks of paper or plastic). However, this applies only to a small quantity of the waste.

In view of what has been said in this section it can therefore be established that a changeover to a 'perfect' levy system for all waste does not solve problems, but rather creates them. The allocation gains that may arise at certain places appear to be more than well balanced by the administration costs that such a reconstruction to a 'perfect' refuse collection levy carries with it.

A limited reconstruction for beverage containers only does not solve the problem of what costs to proceed from. For example, if the cost to be debited to beverage containers is 150 SKr/tonne, the following levies would be applicable:

Returnable bottles: milk (1 litre, presumed weight 400 g)	Approximately 6 öre each (this causes the deposit to increase by the same amount)
Returnable bottles: beer and soft drinks (33 cl, presumed weight 300 g)	Approximately 5 öre each (causes deposit increase)
Non-returnable bottles	Between 2 and 5 öre each, depending on size
Can: juice, beer and soft drinks	Approximately 1 öre each
Paper/plastic: juice, beer and soft drinks	Negligible

(1 SKr = 100 öre, 100 SKr = £10 = $23.) In the selection between different kinds of container, a levy of this size appears to be a certain disadvantage for the non-returnable bottle. The greater part of the large non-returnable bottles (with the highest levy) is used however for wines and spirits where a levy of this size is almost unimportant. This has the total effect that even a reorganization to a 'perfect' levy system limited to include beverage containers alone appears to create problems rather than solve them.

MANUAL SEPARATION IN HOUSEHOLDS

Separation of household waste can involve significant costs to the individual household. Thus, for example, a person living in a tower block could be compelled to go down to the cellar with his separated material, as opposed to the situation today when the same material is thrown down the rubbish chute outside the door. In addition, there are factors such as access to more waste containers inside the dwellings. It must therefore be quite clear that if compulsory separation is increased the total cost for the beverage container whose material is included in the separation will increase. This can in turn lead to a shift to beverage containers not included in the compulsory separation. Decisive for this are the costs the individual household experiences because of the separation and the benefits resulting. It can thus be seen that the cases in which a saving in cost is really provided, because the quantity of waste to be dealt with in the 'usual' way decreases, correspond with the cases (collective living) where it is extremely difficult to check that the compulsory separation is carried out. For one-family houses the opposite applies, as the demand for hygiene requires a high frequency of collection of both unsorted and sorted waste (except when only newspapers and magazines are separated).

At the same time the position is that the volume of separated waste per unit area is relatively speaking much less in areas with detached houses than in tower block areas, which further decreases the possibility of giving the individual household some kind of monetary payment, for example, as a decrease in the total waste collection costs.

Proceeding from this there is therefore reason to ask if there really is any cause to use regulations in order to obtain an increased recovery in addition to that arising under normal economic conditions.

NATIONAL ECONOMIC ASPECTS OF RECOVERY

At present, recovery from household waste takes place to a relatively limited degree in many countries and then mainly as an experimental activity, whilst the reprocessing of industrial waste is carried out to a much greater degree and often has a long history. The amount of recovery is decided here on the criteria of business economy. This section will investigate whether the national economic and business economic bases of assessment can arrive at different decisions depending on, among other things, what is to be recovered and to what extent the recovery should be carried out. Several questions arise.

1. Which factors are relevant from the national economic point of view when recovery is discussed?
2. On which points do differences exist between a business economic and a national economic calculation with regard to recovery?
3. In what different ways can the business economic calculations be adjusted so that from society's point of view desirable levels of recovery can be achieved?

Increased degree of recovery affects most sectors of the national economy. In a national economic cost/benefit calculation on increased recovery the following benefits and costs, with certain assumptions, may apply.

Benefits in connection with recovery:

A. Benefits arise because of the sale of waste resources.
B. When recovery within the country causes a decreased import of raw materials, this may be of value from the point of view of emergency situations.
C. The extraction of natural resources can be decreased.
D. Decreased costs for traditional waste handling (depositing and burning without the recovery of energy) may result.
E. Decreased costs (nuisances) resulting from the decreased environmental stress in different sectors may result, i.e. environmental influences where at present the polluter is not responsible for the cost.

Costs in connection with recovery:

F. various costs (for collection, transport, processing, etc.) in connection with the recycling process,
G. sacrifices of, for example, time and space in sorting in the home,
H. environmental influence from recovery activities.

The extraction of natural resources can be decreased. (Photograph by Bengt Olsson).

A number of the items that may be of interest in a national economic calculation are not taken into account at all or are evaluated differently from the business point of view. Here, two points of view that may be expected will be examined.

Item A on the benefit side in the national economic calculations applies to receipts from sales of waste resources. This benefit is even noted at full value in the business calculations. Thus no difference in relation to the national economic calculation appears here.

The effects in emergencies, item B, are usually not included in the economic business calculations, but may be noted in the national economic ones. The value in this case to be attributed to this item is dependent on the situation in foreign markets, the political situation and the possibilities of substitutes, etc.

The third item on the benefit side concerns the circumstance of a decreased extraction of raw material; although it does not constitute a business profit, it can be evaluated from the national economic viewpoint in a positive way. A decreased 'overconsumption' is felt by many to be very positive.

It is necessary to emphasize, however, that, although decreased extraction of resources is certainly a possible result of increased recycling, the result may just as well be unchanged or increased extraction. In theory, for example, the sorting of paper waste in the household causes a decrease in timber felling and so members of households do not suffer any decrease in forests for recreation. Increased trouble for the household because of sorting can then be said to be compensated by unchanged wooded recreation areas or a landscape with fewer clearings, etc. On the other hand, the reality may in fact be that felling continues quite independently of how much paper Swedish households sort out and hand over for recovery. But the position is still that overconsumption has decreased and, for example, made possible the export of such things as newsprint to developing countries. Apart from the income from exports, the nation's citizens may evaluate this positively.

The possibilities of influencing the actual extraction of natural resources abroad is of course negligible. But even here the knowledge that the overconsumption within the country has decreased through recycling may be regarded as an advantage as the country also contributes to an increased economy of resources in other countries. The feeling of decreasing overconsumption is also a factor that can be included in the national economic calculations, though its real value is considerably more difficult to decide than, for example, export income. Some idea of its importance can be obtained from the waste sorting carried on in certain experimental areas. As the people there cooperate without any more obvious financial gains such as, for example, lower refuse collection levies for the individual household, the conclusion can be drawn that for many this cooperation provides 'benefits' in another form, perhaps together with other things simply the feeling of making a contribution to the decrease of overconsumption of natural resources.

Item D refers to the costs of traditional refuse collection. There is much to suggest that the total national economic costs for refuse collection are now higher than the business costs that the municipalities calculate, as effects on the external environment from dumps and combustion plants, in the shape of aesthetic effects and water and air pollution, are not included in the municipal cost calculations. The decrease in traditional refuse handling costs accounted for in the business calculation thus becomes lower than the comparative item in the national economic calculations if marginal cost pricing is used. Such pricing is not now used in practice so that, in spite of the above, there need not exist any differences between the calculations in this particular respect.

The last benefit item, item E, refers to the costs (inconveniences) caused by environmental effects in various sectors leading to waste disposal. Recovery implies that some of these sectors disappear, with the resultant decreases in costs. In the business calculation this is not entered as an asset as this environmental difference is not today bound up with any business cost.

In the same way, another way of waste handling also means recovery costs for such things as transport and processing. These costs are included in both the business and national economic calculations.

In the cases where recovery implies that separation must be carried out in the individual household, different kinds of cost arise: time and space are required, separation can be regarded as troublesome, etc. Even though these costs do not appear in any business calculation, they still form a national economic cost. That the household takes part in spite of this may, as mentioned earlier, be interpreted to mean that the household thinks it is 'worth the trouble', that is to say, finds the benefits greater than the inconveniences.

Even in this case, recovery activities can of course lead to environmental stress of different kinds. These should be debited to the recovery cost account in the national economic calculations. As the environmental pollution connected with recovery does not constitute any business cost, they are left outside the financial business calculations.

This is summed up in Table 13.2. It is seen that a number of items can be evaluated differently in a business calculation and a national economic calculation. The first, increased preparedness, is not included in the business calculation. The value of preparedness can be given meaning in support for recovery activities of the particular goods or just generally. This support can be interpreted as an insurance premium that society pays to the recovery sector so that it maintains a higher degree of recovery than it would otherwise. In the event of disturbances in imports, the insurance will be paid out in the form of a higher degree of self-sufficiency. One way of formulating this support is to influence the price relationship between imported raw materials (virgin or waste raw materials) and waste raw materials produced domestically, either through the taxation of imported raw materials or by subsidizing waste raw materials produced domestically. So that this direct regulation may be carried out in a correct manner from a national economic point of view, complete information is required about the individual company's production and sales situation. This, however, is not required for a control through the price system (taxes and subsidies).

The other item that can require political adjustment is the decreased extraction of resources — decreased overconsumption. (Here, it is not intended to imply that 'overconsumption' is a reality, but only to suggest how overconsumption can be decreased where it is considered to exist.) The political evaluation of overconsumption can be included in the business calculation, for example, by the price relation between virgin raw materials and waste raw materials being altered to the latter's advantage. If the raw material is imported, the problem is largely the same as when including the item 'increased preparedness' in the business calculation. If raw material (substitute) is both imported and produced domestically, it must be treated in the same way so that different price relations in relation to the waste raw material, having regard to the virgin raw material's origin, are avoided.

As already touched on earlier, there is reason to assume that the national economic costs for dealing with refuse, with regard initially to glass, can be underestimated in a company's financial calculations because of the current levy system. This can in turn be assumed to be accompanied by a lesser involvement in recovery than is motivated by the national economy. An increased recovery

TABLE 13.2 Benefits and Costs because of Increased Recovery in a National Economic and a Business Calculation

Benefits in the national economic calculation	Included in the business calculation
Sale benefits	Yes
Increased preparedness	No
Less 'wastage'*	No
Less traditional waste handling costs	Yes (but can be underestimated)
Less environmental effects in different sectors	No

Costs in the national economic calculation	Included in the business calculation
Increased environmental disturbances	No
Transport and treatment costs	Yes
Separation in the home	No

*Decreased wastage can be justified by the fact that the redistribution between this and coming generations is considered desirable; that is to say, save today for children and grandchildren.

could be made possible here where households are given the task of separating glass from other household waste. This matter will be taken up again later.

On both the benefit and cost sides there is an item covering changes in environmental stress from different sectors of production and use of an article. In certain cases, this means for example that the use of recycled material represents a lesser effect on the environment than the use of new raw material. One result of more stringent demands for environmental care or economic control measures, such as levies for environmental pollution, may be an increased interest in recovery. Thus, recycled raw materials become cheaper in relation to the new raw material if the environmental requirements in the utilization of a new raw material are more stringent. A general measure to express the value set by society on environmental pollution can thus be expected to change the basis of decision in regard to recovery.

The last item on the cost side concerns sorting in the home. This cost is not included in the business calculation, and it is thus quite natural that companies in the recovery sector want increased separation in the home. But in a national economic calculation regard must be paid to the fact that there really are costs for sorting in the home. Some of these costs consist of the actual work of separation and possible aids, such as storage space and sacks. To these must be added the costs for checking that the rules for separation are followed. As a rough measure of the difficulties of checking that separation actually does take place, the size of a house can be utilized. The individual small householder is relatively easy to check. To decide whether household A, B or C has thrown away something 'illegally' in the rubbish chute in a ten-storey block is considerably more difficult, and consequently more expensive.

In a comparison between a national economic and a business calculation, the general result is that, if increased preparedness, decreased overconsumption and decreased environmental damage are considered of interest for society, there is cause to discuss whether it is desirable through regulations, taxes or

Sacrifices of, for example, time and space in sorting in the home. (Photograph by Bengt Olsson).

subsidies, to introduce these evaluations into the business calculation also and in this way support the recovery sector. If, on the other hand, increased preparedness, decreased overconsumption and decreased environmental pollution are not of interest, there is obviously no reason to consider whether compulsory separation should be increased generally, as increased separation means increased costs to the household.

SHOULD BEVERAGE CONTAINERS BE SEPARATED MANUALLY FROM HOUSEHOLD WASTE?

It now becomes of interest to investigate whether any reasons exist to increase compulsory separation to include different beverage containers. Here it may be convenient to presume that a general desire exists for increased recovery to obtain:

1. an increased degree of national independence (the preparedness effect),
2. decreased national consumption in favour of future generations and other countries (overconsumption aspect),
3. decreased environmental pollution (environmental impact).

In making an initial approach with the preparedness effect, it is necessary to investigate what can be manufactured from the separated and collected beverage containers. With present-day techniques it is initially possible to make glass from glass, to make iron rods for reinforced concrete from steel cans, and to extract energy from paper and plastics. Today there is an emergency storage of soda that can be decreased with an increased degree of recovery of glass. For the manufacture of iron bars for reinforcing there is no comparative emergency storage, as there is an abundance of the raw material within the country. There is an emergency store of oil for the production of energy. The position is, however, that an extension of compulsory sorting to include paper and plastic will have the effect that energy must be utilized for the recovery process itself, e.g. for transport, so that the preparedness effect may be modest if it appears at all. It is also necessary to observe that nowadays energy is already extracted from household waste in many countries by means of combustion combined with energy recovery. This recovery of energy occurs in districts with a high population density so that an increased recovery can only be possible in other places without combustion (with energy recycling) which, however, are places with long transport distances between the waste producers.

Concerning the overconsumption aspect, it can be said that for glass it is mainly a question of altered material consumption as the recovery alternative is compared with production proceeding from virgin raw materials. The situation is that various studies about glass have shown that increased recovery does not consume less energy than production from virgin raw materials. Certainly the recovery alternatives mean a decreased utilization of the raw materials sand, salt and lime, but these can hardly be counted among the materials in short supply. The recovery of cans means a decreased usage of iron ore and even this is not in short supply. Additionally, today there already exists magnetic separation of household waste. For the combination paper/plastic it has already been said that the result of decreased 'energy wastage' may be questioned as the recovery itself requires energy.

When finally the environmental impact is concerned, it is also difficult to point out any more marked 'environmental benefits' in the shape of decreased air, water and land pollution.

What can really be concluded quite clearly is that it is only in regard to the recovery of glass that a positive effect may arise, as an increased national independence. In the manufacture of glass, using crushed glass as an ingredient, the saving in soda can be calculated at 200 kg compared with manufacturing 1 tonne of glass from raw materials. 5 tonnes of crushed glass thus correspond to 1 tonne of soda. An emergency storage in the form of crushed glass would, however, be much more expensive than the same amount in the form of soda. Another alternative to emergency storage of soda is the construction of a system for the recovery of materials. In a discussion of such an alternative, recovery should be extended to include other types of glass container, not limited solely to glass beverage containers. If, when and to what degree such recovery becomes realistic should not be decided in advance, but through the introduction of a general compulsory separation. The subsidizing of a recovery system and the corresponding preparedness costs for the storage of soda can be used on the market to decide on the extent of the material recovery.

For other reasons (such as the risk of cuts in handling waste) it may in due course be justified in certain districts and in certain buildings to extend the compulsory separation to include glass, also, which would then include glass beverage containers. In this case, cost reduction may occur. This depends on such things as to what degree the individual household takes part in the separation. The possibilities of refraining from this without risk of any sanction are especially good in multifamily houses. An increase in the disposal costs of glass beverage containers through compulsory separation can therefore produce a highly limited effect. The cost reduction that may occur must be balanced against the cost increase that separation entails because of, for example, increased trouble for the household, increased costs for refuse collection.

Karl Lidgren holds a Ph.D. in Economics and is Professor in Recycling Economics at the University of Lund, Sweden. During the last ten years he has worked as an expert in a number of governmental investigations, principally about environmental and energy questions. As author or co-author, he has participated in the publication of some 20 books in such fields as energy, packaging, the environment and transport.

14 Current Status and Future Prospects for Refuse Management

A. V. Bridgwater

The international meeting on which this book is based brought together experts from 11 European countries to discuss current and future problems of refuse management. At the end of the seminar a workshop examined the evidence and conclusions presented, in order to formulate a consensus view of the problems and possibilities for the future.

It was clear that each country has its own special problems and attitudes which cannot be generalized and fixed to a rigid legislative or political framework. But it was equally clear that much could be learned from others' experiences in both researching and implementing alternative refuse handling and treatment schemes.

Before examining and evaluating the range of methods for refuse management and recycling, it must be appreciated that there is a common and absolute requirement to manage refuse in an acceptable way — i.e. with minimum nuisance, minimum health and safety problems, and minimum environmental impact. Costs are largely irrelevant in meeting this basic objective, although they must be minimized conducive to satisfying these constraints. It must not be forgotten that these standards will change with time as public attitudes change, thereby increasing pressure on governments to tighten legislation. They will also vary with location because of local and natural variations in degrees of legislation enforcement, attitudes and pressures from conflicting interests, and usually the step-change nature of new legislation causing a 'leap frogging' of international standards.

Before attempting to summarize the current situation and then look to the future, it might be useful to very briefly and personally summarize the message of each paper in order to elucidate a theme.

R. Millard. Landfill is seen as the obvious simple and cheap answer to refuse disposal in the short term. However, as landfill costs increase substantially in real terms, as is happening in the UK due to shortage of sites and greater controls, more attention will be directed firstly towards incineration and RDF production for burning, then towards material recovery. Source separation, surprisingly, does not appear as a serious option.

E. Hajek. The packaging industry has changed radically over the last 20 years and, although it now seems to be settling down in some respects, it is in fact very volatile and subject to social pressures. The technology is available for primary and secondary recycling, but the ecomomics are only favourable with co-operation from end users and/or some incentives. Collaboration is necessary between packaging designers, producers, users and industrial consumers; government; recycling and waste management industry; and of course the individual consumer, who ultimately has the responsibility for the success or failure of a recycling principle.

J. Fisher. The possibilities for an established recycling operation in a relatively highly organized and co-operative society are encouraging. The concern remains, however, as to how applicable these might be in different societies with different temperaments, politics, attitudes and social customs. The balance between incentive and legislation is always delicate, even without considering different cultures, and what may be readily acceptable in one country might be almost impossible to attain in others. The relative weights of positive (e.g. fiscal) incentive and legislation compared to other (lesser) forms of encouragement are not well understood and might be deserving of study. Any activities involving individual consumers will suffer therefore from these problems.

J. Paatero. Transport costs are such a significant and wide ranging factor in refuse collection and disposal that considerable care is needed in selecting the optimum. Sufficient research has been carried out to select an optimum for most situations which can have the potential for considerable cost savings. The insensitivity of total disposal cost to minor variations in waste arisings from each household is a notable conclusion.

J. Hansen. The concept of using food waste as an animal foodstuff is very old. With increasing concern about what is eaten, both directly and indirectly, it is not surprising to see attention directed at noxious contaminants of such foodstuffs. Heavy metals are particularly unpleasant as they are so intractable, and it is such problems that seem likely to inhibit derivation of animal foodstuffs from refuse and related wastes, rather than cost or technology. If the source of the heavy metals can be traced, the possibility of reduction or removal exists, but both seem unlikely.

A. Heie. Heavy metals in foodstuffs are a lesser problem in Norway, but there is still concern over environmental health aspects in spite of attractive technology and economics. As a society we are becoming more conscious of our environment and more cautious in setting levels of acceptibility.

J. F. Ingen Housz. We are also becoming an increasingly energy-conscious society, which might be accused of being brainwashed into believing that recycling must be 'better' than a once-through system. This is not necessarily true, as a recycled product such as a plastic may have a different specification than a new product, which significantly affects the apparent energetic advantages of recycling. It is essential to beware of recycling 'fallacies' and avoid being swept along with a recycling tide of opinion without careful assessment of the situation.

A. Buekens. In spite of extensive research and development into mechanical separation systems in North America, there are still some problems in operation which are largely technical and could probably be overcome. There are also problems in selling the recovered materials, which are partly technical, partly economic, and partly attitude, displayed as a reluctance or unwillingness to buy an unconventional material. Most of the problems relate to the heterogeneity inherent in refuse and particularly to unpleasant materials such as mercury batteries, packaging laminates of metal and plastic, chemicals and aerosols. The opposing views of source separation versus post-collection separation will always create discussion, and there are advantages in letting others solve the problems.

O. Tabasaran. It is not difficult to generate substantial quantities of methane from refuse landfill sites — indeed it is difficult not to generate methane, and the problem is how to control the release of the gas to avoid hazards. While a landfill operation may be designed for methane collection, the additional costs of landfill design, gas collection, pumping and piping, and monitoring will almost certainly exceed the value of the recovered gas and except in special circumstances is unlikely to be justified. It may be more viable to encourage methane formation and dissipation in a controlled way, to return the land to valuable use more rapidly and reduce associated hazards.

B. Enhörning. Incineration as a method of disposal has long been practised; now it is common to recover the heat for export in a variety of forms. Incineration with heat recovery for refuse disposal/recycling is now widely used and is usually both technically and economically acceptable as long as there is a ready market for the energy product. Systems are being developed and installed to accept RDF products and incineration with heat recovery appears to offer the brightest prospects, at least in the short term, of viable recovery of values from refuse. With a well established international process contracting industry, there should be few problems of technology transfer through Europe, although local prejudices may delay implementation.

T. Bridgwater. The very extensive worldwide research and development programmes on thermal processing of refuse and related materials from 1969 have resulted in one commercial installation, one semi-commercial installation with associated projects and one substantial demonstration plant — not a good record considering the resources expended. It is difficult to see further commercial development without there being sound commercial justification for the capital sums required, although viability will improve as energy costs increase in real terms. Even the prospect of manufacturing high-value chemical/fuel intermediates such as methanol cannot be considered seriously in the short term. Longer-term prospects with mixed feedstock, including coal, for example, contain some interesting technical and economic opportunities.

K. Lidgren. By taking glass as a case study, the range of possible options for recycling and encouraging recycling are explored with the simple conclusion that the problem is so complex and interactive that there is not a simple answer. The range of fiscal, social and technical incentives are explored in an attempt to establish how best to effect or improve recycling. But perhaps recycling is not necessarily a 'better' option, and resources should be concentrated on recycling those materials for which there are demonstrable advantages.

These somewhat personalized views are included to summarize the current situation from a variety of geographical, institutional, social and technical viewpoints. This appreciation of current achievements and problems in various parts of Europe and the techno-economic possibilities is important in trying to determine how waste management might develop.

At the conclusion of the meeting a workshop was held to examine the current and future problems. These are separately discussed below with particular consideration being given to answering the questions set out in the introductory chapter of the book.

CURRENT STATUS

The key problem of household waste management was identified in the workshop as the need to manage waste in a totally acceptable way. It was accepted that standards vary from country to country and also that these standards will be increasingly raised with time as society becomes more conscious of its environment and standard of living.

Cost is therefore of secondary importance, but must be minimized subject to meeting the basic objective of satisfactory management.

In meeting the basic objective there are a range of alternatives available to be employed individually or in combination — i.e. disposal, treatment, recovery — which may be applied to the whole waste, individual constituents such as metal or glass, or combined constituents such as refuse-derived fuel (RDF).

The constraints against which any waste management alternative must be judged include:

Administration and organization
Amenity standards
Costs
Efficiency
Environmental protection for air, land and water
Fiscal incentives
Legislation
Manufacture of products resulting in waste
Market opportunities
Occupational health and safety
Politics
Product design for maximum recyclability
Product design for minimum wastage
Resource depletion
Sanitary requirements
Specification of recycled products
Stability and reliability of supply
Substitution
Technical feasibility

The alternatives available for disposal (D) and/or recycling (R) include one or, more usually, combinations of the following. Recovery in fact implicitly provides a disposal facility for at least some of the waste.

Digestion		R
Fermentation		R
Incineration	D	
Incineration with heat recovery	D	R
Landfill	D	
Post collection separation		R
Source separation		R
Thermal processing (gasification, pyrolysis)	D	R

(Note: in all cases collection, conveyance and intermediate handling operations for cost effectiveness are necessary included.)

One or more recovery options may be practised, and in all cases a proportion of the material processed will inevitably require disposal. There are thus several hundred combinations of the above operations, employing possibly one or more recycling alternatives with one or more disposal options.

The third and final classification is by those products which may be recycled:

Chemicals
Energy
    heat
    power
    electricity

Foodstuffs
Fuels
  gas
  liquid
  solid (other than RDF)
Glass
  whole containers
  crushed (cullet)
Metal
  whole containers and articles
  shredded mixed metal
  sorted metal fractions
    aluminium
    steel (tinned)
    copper
Paper
Plastic
Refuse-derived fuel (RDF)
  fluff
  densified
Rubber
Textiles
(Ash and inerts)

A complete evaluation of the alternatives, the constraints and the possibilities would occupy a substantial multi-dimensional matrix. A solution was not attempted.

A partial solution may perhaps be found in what has been done, since implementation is probably the best guide to the result of a historical evaluation, but will not necessarily aid determination of the future. Some indicators may however be deduced:

Material and energy recovery is probably independent of satisfying the basic constraints

Detailed analysis, such as cost-benefit analysis on the multi-dimensional evaluation suggested above, does not necessarily give an optimal solution.

Cost appears to be of paramount importance. Since the western world is orientated to a profitability criterion, it could be assumed that the least costly alternative will always be taken, still subject to the basic criterion of acceptable disposal. This premise, however, is concerned only with direct or obvious costs — political, social and such costs are too rarely considered in selecting an alternative for waste management.

CONCLUSIONS

After consideration of the technical, economic, social and and political aspects, the following conclusions are drawn in answer to the questions put in Chapter 1, which formed the justification of this book:

Question 1 *What Cost Savings are Available from Reduced Waste Arisings?*

Money cannot be saved by small reductions in quantities of refuse for collection.

Therefore the economics of different ways of intervention must be economically self-justified without allocating a proportion of the centralized collection and management costs. This arises from the nature of the costs of collection handling and disposal of waste, when total cost is relatively insensitive to modest reductions in quantities of refuse at each household.

Question 2 *Is a Change to Refillable Bottles Always the Best Choice?*

For beverage containers, it is important to distinguish between carbonated and non-carbonated contents. For carbonated drinks, a change from one-way containers to, or maintenance of, refillable bottles or returnable containers (such as aluminium cans with deposit) will respectively cause a reduction, or avoid an increase in waste volume. For some examples of non-carbonated beverages, such as milk, a change to refillable bottles from paper or plastic packaging will, from some points of view, adversely affect the composition of waste. Such a change can cause an increased cost of packaging with handling and transport, a reduced value of waste and overall probably a perceptible increase in costs to the consumer.

It is fallacious and potentially dangerous to assume that recycling for its own sake is necessarily a 'good' course of action. There should be sound justification for implementation of recycling and for provision of any fiscal incentives for encouraging recycling. Or, expressed more simply, 'do the right things for the right reasons'.

Question 3 *Who Should pay for Waste Management?*

The current somewhat negative attitudes to household waste management and the remote methods of financial charges to pay for disposal and treatment preclude direct involvement by producer or consumer. Fiscal incentives to encourage more positive action are discussed in the answers to questions below. These are orientated towards providing industrial solutions, for example by designing more effective disposal or recycling into the product, or by providing a more stable market for recovered materials. Suggestions for introduction of a waste treatment charge for consumer products are superficially attractive, but cannot be regarded as practical or realistic. Any economic reorganization of household waste management is likely therefore to occur in a central administration and involve industrial participation. This does not necessarily apply to capital goods, however, for which in many cases there are already inbuilt financial incentives to the consumer to recycle, such as the scrap value of the metal content. In other cases a waste treatment charge built into the cost of the goods has some merits and is deserving of study.

Question 4 *Should Centralized Sorting Schemes be Installed?*

It would be unwise to invest money too rapidly in centralized sorting schemes: the techniques are unproven and capital intensive, and markets have not yet been developed for the recovered products. Further justification lies in the uncertainty surrounding the future of source separation, which, if implemented, could have a negative effect on the viability of centralized sorting.

Question 5 *What about Fiscal Incentives?*

Fiscal incentives to artificially encourage direct recycling may sometimes have disadvantages since, without a thorough evaluation of the viability of a proposal, direct recycling should not be assumed to be an optimal course of action. Any budgets allocated to encourage recycling might, for example, be better spent on encouraging new product design to facilitate recycling, or development of new technology for utilizing recovered materials.

Question 6 *How can Recycling be Encouraged then?*

In this context, possibly the most useful action is to provide support to a financial base to aid development of viable recycling opportunities, for example with a guaranteed floor price for recovered materials. This would reduce much of the uncertainty associated with the recycling industries and help to provides a more stable industrial climate in which new and existing concerns may operate.

Question 7 *How Important is Energy?*

The current preoccupation with energy content and cost should not be allowed to dominate evaluation of a recovery proposal. Evaluation should be comprehensive and fairly consider all aspect, including economics, technology and social and political constraints.

Question 8 *How Important are Attitudes towards Waste Management?*

Attitudes of both producers and consumers are difficult to change, even with a range of fiscal and social incentives, although these do help. It is uncertain how much can be transferred between countries, and there are dangers in assuming that current practices in one country can be readily accepted in another. It must not be forgotten that both producers and consumers have a responsibility and a role to play in controlling and managing waste more effectively.

Question 9 *Is Material Recovery Preferable to Energy Recovery?*

Energy recovery as heat is already viable in many European situations, and as energy costs increase in real terms is likely to become increasingly attractive. Energy recovery as solid fuel (RDF) is currently the preferred material to be recovered, with ferrous metal as a profitable by-product. Other energy recovery proposals as higher grade gas or liquid fuels are capital intensive and second/third generation technology for the future.

Material recovery proposals have not yet achieved viability in North America and are viewed with scepticism in Europe.

It should not be forgotten that the principal constituent of refuse is paper, board and related materials, without which the energy content would be drastically reduced. Any trends to change the paper/board content of refuse appreciably, or indeed of any other high energy content constituents, need to be carefully monitored. Paper might be viewed as a high-energy content resource which provides a valuable function between the renewable energy resource of biomass and its utilization by combustion to provide thermal energy. The economic and energetic advantages of single and multiple loops to produce paper, recycled paper, re-recycled paper, etc., deserves more careful examination.

The choice between material and energy recovery is dictated mostly by economics — which is the most viable option in the particular situation being evaluated taking cognisance of all the actual and notional costs and benefits? It is important that a thorough evaluation be carried out into each proposal and generalizations are not possible.

Question 10 *Is there a Shortage of Landfill Sites?*

There is no evidence of a shortage of landfill generally in Europe, although there will be an inevitable tendency for suitable sites to become more remote, to have increasingly stringent control measures imposed, and consequently become more costly. As these real costs increase, containment can be aided by more efficient handling procedures, but these must not be developed to the exclusion of other possibilities, which might be even more advantageous and efficient given the same encouragement. As the incentives for recycling increase, a tendency would be expected for recovery of the most valuable materials. This suggests that the least valuable components such as glass, ashes, stones, etc., will still require landfill facilities and these are, in fact, the most suitable landfill materials.

These conclusions not only provide indicators for the future for both disposal and recycling refuse in all forms, but also remind all those concerned with refuse management that a wide range of possibilities exist as alternatives. Pertinent advice and warnings are included to aid evaluation and assist all those involved in planning and policy determination. It is essential that every proposed variation in any declared waste policy be carefully considered from society's viewpoint as well as techno-economic evaluation. This applies to the direct waste problems as well as any secondary effects such as avoidable increased costs to the consumer. Finally from this discussion comes the reassurance and confidence of awareness from a broader understanding of the problems of refuse management.

FUTURE

Cost seems likely to have an overwhelming effect on waste management, at least in the short term. As landfill is still the least costly disposal option generally, it will continue to be preferred both as a total disposal route as well as for the inevitable residues of every other processing alternative. This prediction does not, of course, preclude any recycling operation, but this would have to be economically justified by the public or private sector that would implement the operation.

As landfill costs in real terms increase due to scarcity and remoteness, those alternatives that are currently more costly will become viable, such as incineration. There does, however, seem to be a tendency in some countries to retain landfill and contain costs by more effective transportation systems, when landfill is still claimed to be cheaper at up to 100 miles (150 km). Such a choice may really be cheaper, or it may be a result of inadequate evaluation of alternatives or a result of politico-economic pressures. Nor must it be forgotten that similar intensive optimization of other waste treatment schemes could produce even more significant savings.

As far as materials and energy recovery is concerned, reliance will continue to be placed on commercially profitable operations for implementation of recovery schemes. It must not be forgotten that any recovery operation will reduce the pressure on landfill.

What, then, will change the current situation of only having commercially viable recycling operations?

One major extrinsic factor is the projected increase in real costs of energy. This will enhance the viability of energy (and fuel) recovery schemes such as incineration with heat recovery. As long as there is a positive net energy yield in the recycling process, this will become increasingly attractive as energy costs increase. However, increased energy costs will also affect the quality and quantity of waste arisings in an indeterminate way but probably in a way that will somewhat reduce the attraction.

An even more significant but intrinsic factor is development of an attitude of encouraging recycling. A significant proportion of the cost of recycling is attributable to collection, handling, transportation, etc. If some of these costs were removed by involving the waste producer, in this case the consumer, many recycling proposals would become much more economically, and often technically, attractive. Thus the consumer could separate at source to avoid the necessity of post-collection separation and give a cleaner and possibly more recyclable material. A good example is glass, which cannot be economically recovered from mechanized sorting processes, but which as cullet from a bottle bank is a viable operation, and even more attractive is a whole bottle for re-use. Such attitudes can be encouraged by various incentives such as the goodwill associated with charity collections, or legislation to require source separation.

Finally, and possibly most important, there must be consideration of the materials that are discarded to form refuse. The most significant artificial form, in terms of quantity and value, is packaging. The packaging industry therefore has a major role to play in determining the future of recycling. The present trend to multiple-material, throw-away packaging is hardly conducive to technically and economically efficient recycling. Contrarily, there have been suggestions that, overall, it is more economical to produce low-cost throw-away packaging than to design higher-cost recyclable packages or materials, with additional capital, energy and labour requirements. Packaging design and objectives is clearly a fertile area for investigation in a world of diminishing resources and increasing awareness.

In all the cases discussed above a variety of fiscal incentives can improve the rate of development of change and influence longer-term strategy at the expense of short-term expediency.

ACKNOWLEDGEMENTS

This is a summary of the conference, and a report on the current problems and future possibilities for waste management of refuse as discussed in the concluding workshop. It could not have been completed without the enthusiastic cooperation of all the participants nor without the opportunity for the meeting so excellently arranged by Karl Lidgren.

APPENDIX
1980 Exchange Rate

Country	Currency	1980 average	1980 monthly average minimum	1980 monthly average maximum
Austria	schillings	12.938	12.384	14.000
Belgium	francs	29.243	27.979	31.723
Denmark	kroner	5.6359	5.3828	6.0509
Finland	markkaa	3.7301	3.6200	3.8720
France	francs	4.2260	4.0382	4.5674
Germany	deutschmark	1.8177	1.7236	1.9727
Netherlands	guilders	1.9881	1.9035	2.1419
Norway	kroner	4.9389	4.8183	5.1646
Sweden	kronor	4.2296	4.1258	4.4080
Switzerland	francs	1.6749	1.5929	1.7836
UK	pounds	0.4299	0.3674	0.4270
USA	dollars	1.0000	1.0000	1.0000

The economic data reported in this book refer to early 1980 unless otherwise stated. Exchange rates for all countries mentioned are included in Table A1 with all currencies related to the US dollar. (These are the rf and rh series in International Financial Statistics which provide conversion factors for trade and other flow or average statistics and are period average market rates.)

Average values for 1980 are recommended for approximate conversion, as the range of monthly averages is seen to be considerable and no information is available for precise dates for specific items of information. The monthly minimum and maximum are included to indicate the extent of variation in exchange rates over 1980.

Reproduced from International Financial Statistics.

Index